Thomas R Lombard

The New Honduras

Its situation, resources, opportunities and prospects, concisely stated from recent personal observations

Thomas R Lombard

The New Honduras
Its situation, resources, opportunities and prospects, concisely stated from recent personal observations

ISBN/EAN: 9783337289355

Printed in Europe, USA, Canada, Australia, Japan

Cover: Foto ©Suzi / pixelio.de

More available books at **www.hansebooks.com**

THE NEW HONDURAS:

ITS SITUATION, RESOURCES, OPPORTUNITIES AND PROSPECTS,

CONCISELY STATED FROM RECENT PERSONAL OBSERVATIONS.

BY

THOMAS R. LOMBARD.

BRENT...
CHICAGO.

Entered according to Act of Congress, in the year 1887,
By THOMAS R. LOMBARD,
In the office of the Librarian of Congress, at Washington.

P. F. McBREEN, Printer, 61 Beekman St., New York.

TO THE GOVERNMENT AND PEOPLE OF HONDURAS,

REPRESENTED BY THEIR ILLUSTRIOUS PRESIDENT,

SEÑOR GENERAL DON LUIS BOGRAN,

THIS VOLUME IS AFFECTIONATELY

Dedicated

BY THE AUTHOR,

WHO TAKES THIS OPPORTUNITY OF GRATEFULLY TESTIFYING

TO THE KINDNESS AND HOSPITALITY ALWAYS RECEIVED BY HIM

IN HIS SOJOURNINGS AMONGST THEM.

CONTENTS.

	PAGE
DEDICATION	3
PREFATORY STATEMENT	5
Sources of information.	

CHAPTER I.
INTRODUCTION 7
   Geographical features—Chief cities—Routes and mode of travel.

CHAPTER II.
HISTORICAL OUTLINE 18
   Spanish tyranny—Independence—Civil conflicts—Peace.

CHAPTER III.
CLIMATE.—*Dr. Henry Timm* 26
   The wet and the dry seasons—Natural attractions—Rules of health—Precautionary measures.

CHAPTER IV.
POLITICAL CONDITION.—*Floyd B. Wilson* 35
   Constitutions — Schools — Culture — National progress — Concessions to investors—New roadway—Security of person and property.

CHAPTER V.
AGRICULTURAL RESOURCES AND NATURAL PRODUCTS . . 43
   Cotton—Sugar-cane—Coffee—Fruits—Precious woods—Vegetable fibres—Cattle—Opals.

CHAPTER VI.
THE MINING INDUSTRY.—PAST HISTORY 56
   Placer-mining—Hand-labor—Early records—Population—Council of the Indies—Primitive methods—Mexico.

CHAPTER VII.
THE MINING INDUSTRY.—CONTEMPORARY HISTORY . . 63
   Obstructions—The water line—*Posas*—Ventilation—Petition of 1799—Santa Lucia—Mina Grande—Native workmen—Pay streaks—Smelting—Iron—Quicksilver.

CHAPTER VIII.

GENERAL INFORMATION 73
  Railway projects—Olancho—The Rosario mine—The Animas mine—Lead and fuel—President Bogran's attitude—Heat and disease—Character of the natives—New machinery—Mines not exhausted—Why?—Trade—Gradual development—Exchange—Cost of labor—Outlook.

CHAPTER IX.

YUSCARAN 89
  Policy of the Government—Formation of syndicate—Yuscaran—Its situation, trade, early history—Discovery of mineral wealth—Calvo's adventure—Results—Growth of the camp—Chief veins—Internecine struggles—Bennett's scheme—The great nugget—General desertion—Recent improvements—New awakening to activity—Social features—Concluding statement.

ILLUSTRATIONS.†

NATIVE INDIAN HUT	*Frontispiece*
YUSCARAN .	6
TEGUCIGALPA	18
ROAD-MAKING IN THE MOUNTAINS . . .	34
TUNNEL—SAN ANTONIO MINE . . .	43
SAN ANTONIO VILLAGE	55
SHAFT HOUSE—QUEMASONES MINE—YUSCARAN	62
NATIVE CONCENTRATING VATS	70
NATIVE METHOD—AMALGAMATING ORE, "PATIO PROCESS" .	80
ERECTING MACHINERY AT SHAFT, "GUAYABILLAS MINE," YUSCARAN	88
SAN ANTONIO MINES	100

† The illustrations are from pen-and-ink sketches from photographs taken by myself in Honduras. The frontispiece from a photograph.—T. R. L.

Yuscaran

PREFACE.

CONSIDERABLE interest has been manifested of late in regard to the agricultural and mineral resources of the republic of Honduras. The desire for some plain statement of facts relative to a subject at once so interesting and so important has become a positive want. To meet this want the publishers beg to offer the short summary of information to be found in the succeeding pages. There are extant certain books of travel, exploration, and adventure, assuming to embrace the distinctive characteristics of this particular region within the scope of their pen-pictures. But the familiar *couleur de rose* which is the prevailing tint in these and similar records, renders the works of these writers liable rather to be set aside as mere tales of the traveler, than to engage the serious attention of any persons who might be glad, upon sufficient warrant, to avail themselves of opportunities for the investment of capital in that locality.

It may indeed be remarked that the present volume does not enter largely into details, although the subject presents many and varied features. Fullness of treatment has been subordinated throughout to the requirements of an actual knowledge of the facts in hand. With the object kept clearly in view of furnishing reliable *data* only, care has been taken to eliminate all information of a doubtful character, such as might otherwise have afforded material for expansion.

For the facts from which this book has been compiled, grateful acknowledgment is hereby made to the several

PREFACE.

gentlemen who have kindly furnished the requisite material, gathered by themselves during the time of their personal residence in Honduras, covering a period so recent as to bring the reader almost to the very hour of publication.

So substantial are the peculiar attractions of Honduras to-day, as a field for investment, that they will bear the closest investigation. With the avowed object of encouraging and assisting such investigation, the publishers put forth this book; placing at the same time upon record a declaration of their confidence that with the full attainment of this object, the new Honduras will emerge from the obscurity in which she has so long lain dormant. For the dissemination of an intelligent understanding of the opportunities she offers will result in a practical appreciation of them, which can have no other issue than the gradual and successful development of her great natural resources.

Among those who have assisted in this work, both in writing the special chapters credited to them, and in furnishing valuable *data* from which other chapters were compiled, the editor desires particularly to name Floyd B. Wilson, Esq., Henry Timm, M. D., and Messrs. Henry Leeds, Jr., Thomas J. Foster and George S. Evans.

THE NEW HONDURAS.

CHAPTER I.

INTRODUCTION.

A DAY'S journey in Honduras will disappoint one who expects to find there the luxuriant growth of vegetation commonly supposed to adorn tropical and semi-tropical countries. Excepting along the river-banks near the coast, the vegetation partakes rather of the orderly character of the growths of the temperate zones than of the rank and noxious jungle barely penetrable by the rays of the sun.

Once past the mahogany belt, covering the low land for twenty miles on the south coast and forty on the north, straightway we rise up among high plateaus and mountain valleys. Here the temperature is low enough for the pine to flourish, and yet so equable as to encourage the countless orchids which well nigh clothe the hillsides, as well as the abundant *cacti* and an endless variety of palms.

The fact that Honduras does not present the exuberant characteristics only too often figuring in vivid descriptions as the offspring of an artistic fancy, is due to the geological formation of the country rather than to its latitude.

Geographical Features.—Lying between latitudes 13° and 16° north, and longitudes 83° and 90° west, it is in reality less tropical than some portions of Mexico situated further towards the north. This is readily accounted for by the comparatively narrow strip of swamp-land along either coast, and by the rapid rise of the surface of the country to an altitude precluding the existence of a humid soil.

The country has an area of 48,000 square miles, or about that of the State of Ohio. In its greatest length from east to west, *i.e.*, from Cape Gracias-a-Dios on the Atlantic, to the Cerro de Biujo on the boundary line of Guatemala, it covers a distance of 388 miles. The width, from the north coast to the Pacific, is about 175 miles as the crow flies. To the north lies the Bay of Honduras and the broad Atlantic; eastward the republic of Nicaragua; to the south, the Bay of Fonseca and San Salvador; while the thriving republic of Guatemala lies conveniently adjacent on the west.

Honduras is a table-land. Its topography shows a series of elevated plateaus, broad savannas and mountain ridges. Some peaks of the latter rise to an altitude of 8,000 feet or more, above sea level. The principal range of the Cordilleras, running east and west, marks off the two great divisions of the country. Of these, the wider extent of territory and the larger districts devoted to agriculture and cattle-raising belong to the northern section. On this side of the "divide" are the wide plains of Olancho and Yoro, and the fertile valleys and uplands of Comayagua and Santa Barbara, heavily timbered and rich in grasses and shrubbery.

Innumerable streams water these districts, navigable only in exceptional cases and for short distances. The Guayape, draining the great Olancho country; the Ulua, performing a like service for Santa Barbara; and the Aguan, rising in the mountains of Yoro and finding its outlet eastward of Truxillo as the Roman River. These streams are navigable only to the verge of the mountainous district; yet in their windings they traverse even thus far a considerable territory.

South of the Cordillera range, the one river of special importance is the Choluteca, receiving the contributions of several smaller streams from the mountains of Choluteca and Tegucigalpa. It has a southerly course to the Bay of Fonseca, south of the island of Tigre. Honduras has jurisdiction over the Bay islands off the north coast, as well as

INTRODUCTION. 9

over the islands of Tigre and Sacate Grande, in the Bay of Fonseca.

In this southern section the principal cities are Tegucigalpa, Comayagua, Juticalpa, Santa Barbara, Santa Rosa, San Pedro Sula, Truxillo, Yuscaran and Amapala. These points are the centres of trade and commerce for the country.

The first mentioned, Tegucigalpa, is about 3,000 feet above sea level, at the junction of the Rio Chiquito and the Rio Grande. Its population is estimated at 12,000.

Comayagua is advantageously located on a level plain, and has at present some 8,000 inhabitants. Old chroniclers assign to this city, which was founded by the Spanish adventurers, a population of no less than 40,000. At an early period of its history, it was undoubtedly an important station midway between the oceans, and far more populous than it is to-day. Near it are two rivers, the Comayagua and the Chiquito. The centre of trade in the western part of this section is Santa Rosa; and in the Olancho district, Juticalpa, situated on the Guayape. These places have each about 4,000 inhabitants.

In mining affairs, the present seat of activity is the town of Yuscaran, situated on the slope of a mountain overlooking a wide and pleasant valley. It is the capital of the Department of Paraiso, and its people are mainly employed in the mining industry.

San Pedro Sula derives its importance from its nearness to Puerto Cortez, some forty miles distant. From Puerto Cortez great quantities of fruit are annually shipped to the United States. San Pedro is situated on the plain of Sula, at the foot of the mountain range of Santa Barbara. It is a dépôt of supplies for the northern and western part of the country. One of the chief ports of entry is Truxillo, overlooking the little Bay of Truxillo, formed by the projection of Point Castilla into the Atlantic. Hence are exported the products of this western region. The town of Amapala is the only sea-port of Honduras on the Pacific side. It occupies a commanding position on the island of Tigre, at the base of a low extinct volcano, and overlooking the beautiful Bay of Fonseca.

This bay is beyond question one of the finest ports on the entire Pacific coast of the American continent. It is upwards of fifty miles long, by thirty in average width. Its entrance from the sea is about eighteen miles wide, between two towering mountains. Several large islands stand as though moored within this mighty harbor, completing the picturesque character of the scene, from whatever point of view the traveler beholds it. A direct trade is carried on between Amapala and the chief commercial ports of Europe and America. The bay abounds in fish, and its shores swarm with every variety of water-fowl,—cranes, herons, pelicans, ducks, curlews, &c. Large beds of oysters are found in the shallow waters and in the dependent bays. Their quantity seems to be inexhaustible. Huge piles of their shells are scattered along the shores of the islands and mainland, showing in what appreciation they are held by the natives. These oysters are about the size of the ordinary variety found around New York, and of excellent flavor. Crabs and lobsters are also abundant.

The island of Tigre, a huge lofty cone, is the most important in the bay. It was long a favorite resort of pirates. Here it was that Drake had his dépôt, during his famous operations in the South Sea.

These smaller cities of Honduras vary somewhat in commercial importance, but the trade they foster furnishes occupation and sustenance in each case to a population of several thousands.

Political Divisions.—Of these there are eleven, known as the Departments of Tegucigalpa, Comayagua,

INTRODUCTION. 11

Olancho, Colon, Choluteca, Santa Barbara, Copan, Gracias, Paraiso, La Paz and Intibuca. The most important as regards population, mineral wealth, and agricultural resources are, Tegucigalpa, Paraiso and Olancho. Some additional information concerning these three departments will therefore be of interest to the reader.

The department of Tegucigalpa has undergone subdivision of late, but is still the third in size and perhaps the most important of the divisions of the republic. It includes the commercial centre and seat of government of Honduras, and some of the most famous mineral districts of Central America. The population of the capital city is given above. No census has as yet been taken of the whole province. The interior of this department is a plateau known as El Portero, where the capital is located, enclosed by mountains rising from two to three thousand feet above its own level. Here the climate is remarkably cool and salubrious, being compared with that of a perennial spring-time.

The name Tegucigalpa is a Spanish corruption of an Indian word signifying "mountains of silver;" and the hills around the city have yielded precious metal in such abundance as to warrant the appellation. Here are the *minerals* of Santa Lucia, San Antonio, Los Angeles, Guasucaran, Villa Nueva, El Plomo, Cedros, San Juan de Cantaranas, Santa Anna, and Barrajanas.

The chief resources of this region are mineral, the agricultural advantages not being so marked as in other sections of the country elsewhere described. According to official statistical records, kept on file among the archives of the city, the mines of the adjacent territory have yielded immense quantities of gold and silver, and that too under the disadvantages of the crudest possible systems of mining and processes of reduction. Indeed it is believed that in no other known region of equal territorial extent are to be found so many veins bearing the precious metals. The principal qualities at present extracted are those ores known as sulphuret and galena. It is true that to-day the native workmen produce but comparatively insignificant results, but this fact is owing to their primitive mode of operation and desultory efforts. That there is an abundance of material ready to hand, and only awaiting proper methods of mining and reduction to encourage the generous response of former days, does not admit of a doubt.

For example, seven miles north of Tegucigalpa lies the Santa Lucia mine, one of the oldest and in time past one of the most richly productive districts of the whole country. The amount of wealth received in the form of royalties from this region by the crown of Spain, seems almost fabulous, although the facts are in the records.

Though scarcely a league square, Santa Lucia is known to contain no less than two hundred distinct veins of ore. It is bisected by the Rio Chiquito, whose banks show the twofold rock formation of this particular region ; the southern side being characterized by limestone and the northern by porphyry. This river provides abundant water-power in close proximity to the mines. The village or *pueblo* of Santa Lucia itself is most picturesquely situated. It bears unmistakable evidence of having been founded in the expectation that it would become a permanent centre of mining operations. The streets are paved and terraced, and the white washed adobe houses shine brightly in the sun, when viewed from the adjacent hills. An old aqueduct, built by the Spaniards, still furnishes the town with its water-supply for all ordinary purposes.

Twenty miles beyond is the Los Angeles district, where similar conditions exist. About the same distance to the southwest is the San Antonio mine, with two settlements, one on either side of a ridge containing the ore-body. In both these *minerals*, large quantities of ore are in sight, chiefly of the kind known as argentiferous galena. The veins of Tegucigalpa vary in width from one yard to ten or twelve yards. The celebrated "blue vein" of Cedros runs as high as forty-five to sixty yards in some places; but this is far above the average.

Adjoining this department and formerly included within its limits is that of Paraiso. Here, within half a league of the town of Yuscaran, there are thirty-five known gold and silver mines.† Next in importance is the department of Olancho, about as large as the State of Maryland, and famous in Central America as a cattle district; while everywhere its streams and alluvial deposits are marked with *placer* gold.

In the dry season of the year, considerable quantities of gold are washed from the streams by the Indian women, who by the use of the primitive "pan system," are accustomed to make their living in this manner, alternately working and idling. The river bed of Guayape has been famous for its yield of gold-bearing sands, from the landing of Columbus down to the present day. The average depth of this stream is three feet in the dry season. During nine months of the year the depth is greater, and such obstructions to navigation as exist might readily be removed. These consist merely of rocks of no great size, which divide the course into narrow channels at various points. As their removal will open up one of the best natural water-ways from the interior to the coast, there is good reason to believe that the government will not long delay the accomplishment of this improvement, which it has had under consideration for some time.

The inhabitants of Olancho are remarkable for sturdiness of character and an intense love of locality. They have kept themselves somewhat aloof from the general government, although rendering it all due and proper acknowledgment. Restive under political exactions, and impatient of control in matters concerning home affairs, they are yet,

† See Chapter IX.

14   THE NEW HONDURAS.

in all their dealings with those who seek to elevate the condition of country and people, hospitable to a remarkable degree.

Cattle-raising is one of the chief industries of this section. The department abounds in broad savannas, whose rich soil, well watered by numerous streams, produces luxuriant grasses; while the climate is particularly mild and salubrious. Aside from the valuable woods obtained from the forests, as shown in the chapter on natural products,† it has been practically demonstrated that the soil of Olancho is capable of nourishing nearly all the ordinary products of more northerly climates. Six streams cross the plain of La Concordia, affording sufficient water for hydraulic work in the *placer* regions of that vicinity. Timber in plenty is available near at hand for the construction of suitable buildings and mechanical appliances. A considerable outlay, however, would be required for the establishment of works for the extraction and reduction of gold in large quantities.

Colon and Yoro are most noted for fruits and valuable woods. A greater amount of capital has been engaged in these industries than has ever been expended on the mineral prospects of the entire country. Comayagua and Santa Barbara are given over to a somewhat primitive system of agriculture. The soil of their plains is remarkably fertile in vegetable products. Both have an abundance of veins, bearing gold and silver as well as baser metals, to which little attention has been paid. Copan is devoted mainly to the cultivation of tobacco; and Gracias is renowned for its opals, which rank with those of Hungary.

Routes of Travel.—There are three recognized routes to Honduras from the United States. The first is *via* the Pacific Mail Steamship Company's line from New York to Aspinwall, thence by rail to Panama, and by the

† See Chapter V.

INTRODUCTION.   15

Central American Coast Line Steamers of the Pacific Mail Company to the Port of Ampala on the Pacific side. This journey requires fifteen days, and one first-class fare through is $165 in gold. The other routes are from New Orleans. The Messrs. Mecheca Bros., of that city, run three steamers monthly to Balize (British Honduras), and thence to Puerto Cortez. This line carries the British mail and furnishes a regular and efficient service. The third line is run by Oteri Brothers, whose steamers ply between New Orleans, the Bay Islands, and the Port of Truxillo. From New Orleans, by either of these latter routes, the fare is $35 in gold, and the trip occupies in each case from four to six days. A word of praise is due to the management of these steamship companies for the well-ordered service and superior accommodations given to the traveling public. Indeed it would not be amiss to speak in complimentary terms of the experience, skill and courtesy of their commanding officers.

The ocean journey, of course, is one of ease and comfort. Not until one is landed in Honduras does he encounter some of the difficulties of travel. Prior to the current year a roadway from the coast was a thing talked of merely. The fact that such a road now exists is owing to the enterprise and perseverance of the several American mining companies now operating on the Pacific slope. With the single exception of this wagon-road, all points in the country are connected by mule-trails, taking the most direct course practicable from place to place. These lead up hill and down dale without regard to convenience, and are in many portions not without dangers.

The wagon-road from the Pacific coast is one hundred and twenty-two miles long, and as yet is traveled only by wagons engaged in transporting heavy mining machinery from San Lorenzo on the coast to Yuscaran by way of the capital. Aside from this, freight transportation is still carried on by the pack-mule system, and the traveler there must spend the better part of the day in the saddle.

Although Tegucigalpa is but ninety miles from the Pacific coast, it is a three days' journey, ten leagues being the average distance covered in a day. The cost of transportation between these points by a pack-animal is $6 for every two hundred and fifty pounds. A riding-animal may be hired for the trip for $8, with an additional $8 for the *mozo* or servant, who looks after the comfort of the traveler, and the feeding and care of the animal.

Entering the country from Puerto Cortez, on the opposite side, the Interoceanic Railway carries the traveler from that port to San Pedro Sula, a distance of thirty-eight miles, for the sum of $3 in Honduras money. With a little delay and a great deal of patience, a tolerable riding-animal may be secured at this point for the journey into the interior. The cost of a mule for riding, from San Pedro to Tegucigalpa, a journey of a week, ranges from $15 to $20, according to the season of the year, with a like amount for the servant. Cargo-mules, carrying two hundred pounds, necessitate an outlay of $12 each for the same journey. The cost of provisions for the servant and for the animal does not exceed seventy-five cents per day for both, or a dollar when the servant also has a riding-mule.

The tour from Puerto Cortez to Tegucigalpa leads through the departments of Santa Barbara, Yoro, Comayagua, and Tegucigalpa, a distance of about 250 miles. Though occupying over a week, this trip is one to be recommended, affording as it does a favorable opportunity to study the country and the people. The tourist sees bananas ripening for the northern market. There too are the mahogany forests and cuttings, the sugar and coffee plantations, and the native mining and reduction industries.

By way of Truxillo, the other Atlantic port, ten days on mule-back are required to reach Tegucigalpa, at a cost

INTRODUCTION.   17

of from $15 to $25 each for mules and for attendants. This trail runs through the plains of Yoro and Olancho, the great pasture-lands. It leads also through the *placer* regions, which from time to time attract, quite as often as otherwise, however, to their ultimate regret, adventurous miners from our far western States.

NOTE.—For the benefit of those who may hereafter travel in this region, the following recommendations are supplied from Parker's Guide to Guatemala and Honduras:

"New-comers are advised to provide themselves with a rubber 'poncho' or riding-cloak, riding-boots, cork helmet, hammock, a woolen blanket and one of rubber, spurs, riding-whip, two long towels to serve as protectors to neck and shoulders when the sun is high, a bottle of ammonia, *no cumbersome clothing*, a small cooking outfit,—consisting of a methylated spirit-lamp and the requisite utensils,—and some canned edibles.

Travelers are cautioned against tasting all waters found *en route*, and advised to drink cautiously at first, and to confine themselves to drinking at morning and at evening. Riding should cease at 11 A. M. for an hour and a half. There is little danger in traveling throughout the length and breadth of these countries—the worst danger being from an habitual use of stimulants. These, it should be added, are wholly inadmissible."

CHAPTER II.

HISTORICAL OUTLINE.

IN various parts of the republic of Honduras are found numerous evidences of the early existence of a people whose identity has baffled all attempts at discovery, and even whose mode of life and condition of civilization can only be conjectured.

On some of the sculptured stones of Copan may be traced forms and characters bearing a striking similarity to the venerable relics of Eastern continents, but the bare fact of such resemblance furnishes no satisfactory clew to the mystery. Nothing has ever been definitely determined as to the people who inhabited that country centuries ago, long before the advent of the Spaniard. Their hieroglyphics, which adorn the ruined altars and massive idols of western Honduras, are their only extant records, and these have not as yet found a translator.

It is true that the Castilian adventurers, who overran the country some three hundred and fifty years ago, reported this region as being very populous. Their historians refer to many tribes of the interior, and speak of them as at least numerically powerful. All such statistical opinions, however, must be taken with a large grain of allowance, leaving a wide margin for their well-known liability to exaggeration.

From an incident connected with the fourth voyage of Columbus, Honduras derived its name. It is related that after leaving Point Casinas, the great discoverer sailed eastward toward Darien, encountering many severe storms, and being unable to find any anchorage until he reached the

TEGUCIGALPA.

point now known as Cape Gracias-a-Dios. Such was the feeling of thankfulness and relief after making successful soundings, that the spontaneous exclamation, "Thank God, we have passed those *Hondura* (deep waters)!" became memorable, and took permanence in the name of the Bay and of the country.

Seven years after came Cortez, the accounts of whose expeditions would furnish many a theme for romance. The famous march of this indomitable leader through the untracked wilderness of Mexico and Guatemala, in order to arrest and punish a refractory lieutenant, takes high rank among the achievements of explorers. A very limited experience of travel through the region of that journey of his, would suffice to convince any one of the magnitude and significance of such an undertaking. If true, as we have every reason to believe, then indeed is truth stranger than fiction.

Out of the chain of fortresses established by the Spanish conquerors for the accomplishment of their plan of subjugation, grew the quiet settlements which have developed into the villages and towns of a later and more peaceful period. The intervals of space between them, from the coast to the interior, indicate with strongest probability the truth of this theory of their origin. One of the oldest towns of the interior is said to be Comayagua, very early established as a half-way station between the coasts. The legend of its foundation runs as follows: Once, when the Spaniards, closely pursued by a horde of native warriors, sought long and wearily a place of refuge, they came at last to a somewhat sheltered spot. Here the leaders commanded a halt, in order that the band of fugitives might pause to refresh themselves with food and drink. Having fortified the inner man, they took courage, and built a fortress for defence. Hence the place acquired the name it still retains, the meaning of which is, "Here we ate and drank."

According to the old Spanish tales, every advance attempted in the gradual occupation of this territory was resisted by force of arms. There are, however, but one or two recorded cases in which the foreigner was the sufferer, and in these instances the punishment was well deserved, and tardily inflicted.

Indeed, the history of Honduras from the time of its discovery by Columbus in 1502, until the declaration of its independence in 1821, may be summarized in the statement that it was a period of continual oppression and occasional massacre of the natives by the Spanish settlers. Inhumanity follows inhumanity, until the story of the conquest and settlement of the country becomes only another chapter in the record of Spanish adventure on the American continent. The struggle for gold, and the erection of what was practically a feudal system, all but exterminated the Indian race. The policy of the famous "Council of the Indies," whose jurisdiction included the colonies of Spain, from the latter part of the seventeenth century to that of the eighteenth, tended to retard rather than to further the development of the resources of the country.

A vivid description of the atrocities perpetrated by the Spanish rulers upon the helpless Indian, is given in a letter to Charles V. by a missionary of the sixteenth century, Bartolome de las Casas. His estimate of the population is probably an exaggerated one; but of the cruel practices of his countrymen he undoubtedly had personal knowledge.

The condition of slavery to which the unhappy Indians were reduced, appears from his account to have been the most distressing imaginable. They were compelled not only to extract the precious metals from the earth and to supply their conquerors with luxuries, but to serve in place of mules and horses in carrying tremendous burdens. It is related of the redoubtable Diego de Velasco that he slaughtered over 10,000 of them in a single month. So great indeed were the cruelties practiced upon the natives, that the remnant fled at last from their brutal task-masters, and betook themselves to the mountains, leaving the Spaniards to their own resources and in a state of destitution bordering on famine. Thus runs the record of Las Casas.

From the discovery of this territory until the independence, no step was left untaken by the Council of the Indies to restrain all efforts looking to the intellectual advancement of the natives, or even, for that matter, of the descendants of the conquerors themselves. This Council ruled all the Spanish colonies, and in so doing endeavored to keep the inhabitants in a state of ignorance, the better by this means to control the wealth of the country. Only a few years prior to the independence did the people awaken to the fact that the rest of the world was not dominated by Spain and under tribute to its crown.

In order to make the province more dependent upon the mother-country, the cultivation of any product of the soil of Spain, and the manufacture of any commodities made in Spain, were prohibited by law. For this reason the vineyards planted on the north coast of the province in the first year of settlement, and producing a superior wine, were afterward abandoned. To keep the inhabitants in a helpless condition, resulting from continued ignorance, any intercourse with foreigners or relationship to neighboring Spanish colonies, for trading or other purposes, was declared not only criminal but capital. All books were forbidden by the Inquisition, except the catechism and the prayer book. As to any knowledge of the history of the early conquest of the country, the people were kept as far as possible in the dark.

The idea of an independence apparently took its rise not so much from this extreme tyranny as from the levying of excessive contributions to aid the mother-country in her desperate warfare at home. Such contributions were exacted from rich and poor alike, and served for once to unite master and slave in a common cause against a common adversary.

In the year 1812, several insurrections took place in Salvador and Nicaragua. These were summarily put down. It was not until after the success of the patriotic cause in Mexico that the separation of the Spanish provinces from the mother-country was finally accomplished. On the 15th day of September, 1821, Honduras declared its independence and assumed the position of a sovereign State. Since that time, until the year 1876, the republic has been more or less disturbed by the constant warring of political factions. During this latter period of fifty-five years, Honduras acquired the reputation of a revolutionary republic.

In justification of these more recent disturbances, it must be said that they were the result of a struggle for political principles. The predominating desire has been either for the political union of the five Central American States or for their individual independence. The final separation from Spain had been accomplished without difficulty. It was soon followed, however, by conflicts between the several States, caused by the efforts of ambitious politicians to attain personal control of the general government.

These dark pages of the country's history are illumined by a galaxy of heroes, whose exploits, if done in other lands, would have won for them imperishable glory in the annals of the world. Chief among these is Morazan, the "Washington" of Honduras. Wise in counsel, a most humane and energetic military chieftain, his immortal fame is one of the proudest boasts of Central America, the historic field of his heroic and self-sacrificing achievements.

For ten years past, the country has enjoyed an era of peace, enabling the people to turn their attention to the advancement of their material interests. A complete change of spirit is manifest among them ; and, while pronounced

HISTORICAL OUTLINE. 23

political opinions are still retained, changes are now wrought at the ballot-box instead of at the point of the bayonet. Under the efficient government that now controls the affairs of Honduras, marked advantages are rapidly accruing to the country. Compulsory education has been established. A constitution modeled upon that of the United States has been put in force and is being lived up to with admirable strictness. Special effort is constantly made for the development of national interests. For co-operation in furthering these objects, exceptional advantages are offered. The people realize that they are not sufficiently advanced in the arts and sciences to enable them to achieve, without the aid of foreign capital, that position of wealth and independence which the natural resources of their country guarantee, and to which they zealously aspire. As a consequence, they are to-day more ready to welcome the invading forces of enlightenment and progress than their predecessors were of old to repel the encroachments of the Spanish adventurer.

It has been truly said of Honduras that nature has done everything for the country, and man has done nothing. For this no blame should attach to the people. From a condition of utter ignorance, to be suddenly uplifted by the longings and ambitions of free men, would render them only the easy prey of unscrupulous demagogues. Thus their energies and resources have been frittered away in futile endeavors to decide abstract political questions. But the last few years have seen such rapid advances that a new character, pregnant with future promise, seems to have developed among them. As a people, they are mild in disposition, and upright in character. Possessing much native shrewdness, they quickly become intelligent workmen when their labors are skillfully directed ; while a prevailing docility of temperament ranks them among the most tractable of employees. True, their labor at present commands comparatively small wages, but this is due rather to the simplicity of their wants than to any inferiority in their work. When treated with firmness and consideration, they are quite as amenable to discipline and to the requirements of orderly living as any other class of workmen.

The Honduranians of the present day may be divided into four classes, according to their race-origin. The Castilians, who trace their line to the Spaniard ; the Indians, antedating the conqueror ; the Half-breeds, including the "cross" either with the Spaniard or with the negro ; and those in whose veins flows the blood of all three races. In features, the Honduranians are rather prepossessing ; in form, as a rule, symmetrical and graceful. When rightly treated they are faithful in the discharge of their duty. In fact, more than one traveler has observed, that whatever their lack of progress in the finer arts of modern civilization, they are still happily unlearned in the sophistries that have filled the records of that civilization with instances of broken faith and violated trust.

A further peculiarity is the marked difference between the interior and the coast region, but especially the north coast, observable in the character of the native populace. In the north are found a large proportion of "Caribs," as they are called—a race of blacks believed to have descended from the negroes originally transported thither as slaves. These people display all the familiar characteristics of the African race, a correspondence extending even to the matter of their religious observances. They live in towns and villages where the inhabitants are almost exclusively of their own race, preferring to mingle but little with either the Spanish element or the native Indian. They are accustomed to make their living by working on the fruit farms, by fishing, and by mahogany cutting. It is seldom that they care to go into the interior, except it be in the capacity of servants to European or American travelers. They make good servants, being tidy, obliging, intelligent, and full of clever expedients in cases of emergency. Their language is a dialect of their own, although they can converse in Spanish, as well as in a kind of Macaronic lingo, or pigeon-English, very difficult at first for an American to understand.

In our own day, in short, this whole region rejoices in a revival of general prosperity, which bids fair to issue in the most desirable results for the country and its inhabitants. The coast lines are still sparsely settled, but the territory adjacent to the lowlands nevertheless presents most favorable opportunities to the enterprising and energetic farmer and cattle-raiser. From the boundary-line of Guatemala to that of Nicaragua, there are about four hundred miles of coast, in many portions as yet unexplored save by the nomadic Indian tribes, known as the Wacos or Mosquitos, scattered along the eastern shores. Several navigable rivers offer rich returns to those who in future shall have the good fortune to open up to the world of commerce the forests of rosewood and mahogany, the ungrazed pasture lands, and the auriferous mountain-streams whose treasures only await appropriation.

Some results of the recent impulse given by the local government to business schemes during the last few years are discernible already, and may be noted here. On the northcoast have been successfully established two large concerns, one for lumbering and the other for fruit-raising. The mining companies operating in the interior of the country are under the management of Englishmen, Frenchmen and Americans,—the latter predominating. Every commercial project thus far undertaken, with a judicious combination of capital and skill, has been eminently successful. The liberality of the legal regulations governing the general transactions of trade, at home and abroad, confers upon this locality the strongest attractiveness to all who are interested in the development of rich but latent natural resources.

CHAPTER III.

THE CLIMATE.

TO know the latitude and topography of a country is to know something of its climate. The conditions of the latter are modified by the configuration of the adjacent territory as well as by the chemistry of the soil.

Lying between the 13° and 16° of north latitude, Honduras is near enough to the equator to have a certain degree of tropical heat. Being, however, but 200 miles in width with one coast open to the north-east trade-winds of the Atlantic, and the other saturated by the moist breezes of the Pacific, the climate is rather to be called semi-tropical. The higher table-lands enjoy the climate of the temperate zone. The surface is broken by numerous mountain-ranges, which, with the valleys, savannas, and table-lands, afford every variety of climate. The heat on the Pacific coast is not as oppressive as on the Atlantic; less, perhaps, on account of any marked difference in temperature than on account of the greater dryness and clearness of the atmosphere.

The north-east trade-winds which sweep over the Atlantic reach the continent saturated with vapor, and pass over the whole of Honduras, leaving the atmosphere moist and invigorating. When these winds are heavily charged with moisture they are intercepted by the mountain centres, the vapor is precipitated from the clouds and flows back to the Atlantic through numerous streams and rivers. These mountains are from 6,000 to 8,000 feet high, and are covered with extensive forests, through which sweep the pure soft ocean breezes, giving an atmosphere that is as invigorating as it is delightful. In the low lands on the coast the fever-germs generate, giving this region a somewhat miasmatic character. It is not, however, any more marked by malaria than the marshes of Indiana or the bottom-lands of Kansas. So keen an observer as Charles Dickens found little else than fever, ague, chills and living skeletons in a region which is now rapidly becoming one of the most productive and prosperous sections of the United States. Thus many travelers who visit only the coasts of Honduras, bring back reports that tend to discourage tourists. And this, too, of a country which had many flourishing cities as far back as 1540, and which still maintains a hardy and vigorous people.

There are, speaking generally, but two climatic seasons in Central America; the rainy season, corresponding to our summer and autumn; and the dry season, corresponding to our winter and spring. The rainy season begins about the middle of May. The rains are at first intermittent, and gradually increase until the maximum is reached in July or August, when rain falls every day; then they gradually diminish, ceasing entirely in November. These showers come up about two o'clock in the afternoon and last until five or six o'clock. They are often preceded by strong winds, with thunder and almost continuous lightning. Sometimes these storms last all night, when the roads become heavy, and the streams so swollen that fording is difficult. Vegetation shows new life, and the flora with their wealth of brilliant tropical colors are all fresh and smiling after their bath.

The difference of temperature between day and night is notable. Owing to constant evaporation from the soil along the coasts, rivers and lakes, the atmosphere becomes moist and chill through condensation. The dry season follows, usually from January to April, when little if any rain falls. This is the most convenient time for travelers to visit the Central American countries. True, it is with us

28   THE NEW HONDURAS.

the winter season, but the winter season of Honduras is in reality its summer, since the thermometer averages a few degrees higher than it does during the wet months corresponding to our summer. Even when the thermometer reaches its highest point, the heat is not oppressive, and, the air being dry, the nights are refreshingly cool. In the elevated regions of the interior, on the table-lands, and on the crests of the Cordilleras, we find a mild even climate with luxuriant tropical surroundings.

The mountains which rise around the fertile valleys are ascended by terraces, crowned with forests of pine and oak and carpeted with grass. The summits of the mountains sometimes run up in peaks, but generally constitute broad table-lands, more or less undulating, and often spreading out in rolling fields, traversed by low ridges of vendure, and yellow belts of trees which droop over streams as clear and as cool as those of New England. Here the familiar blackberry is native to the soil, and the bushes which impede the traveler are covered with fruit in its season. Fields of grain, billowing beneath cool mountain breezes, and orchards of peach trees, struggling against man's neglect, give to these districts all the natural aspects of the temperate zone.

Up on the higher mountain crests, where the short and hardy grass betokens a temperature too low for luxuriant vegetation of any kind, the pines and gnarled oaks are draped in a sober mantle of long gray moss, which waves to and fro in the passing wind, like frayed and dusty banners from the walls of old cathedrals. The very rocks themselves are browned with mosses, and except the bright springs gushing from beneath them here and there, and trickling away with a musical murmur, there is no sound to break the stillness.

On the northern coast, the mountains and hills are more diversified with verdure, and there is a greater variety of trees. Cliffs and rocky crests are few, the forests dense, and full of multitudinous forms of animal life.

The average temperature of Tegucigalpa, Comayagua, Juticalpa and Gracias, the principal towns, is 74°. In the plain of Comayagua, situated in the very centre of Honduras, and equidistant from the two great seas, more or less rain falls during every month in the year. During the dry season, however, on the Pacific slope, it appears in the form of showers of brief duration, while during the wet season the rains are long and heavy. The temperature of the Pacific slope ranges from 70° to 95°; the highest being from 12 o'clock noon to 3 o'clock in the afternoon, and the lowest from four to six in the morning. The climate, in general, is healthful. The strip of low marshy land along the coast does not in any way render the country liable to epidemics, providing dietetic and hygienic laws are properly observed.

On the Pacific coast along Fonseca Bay, the traveler comes upon a table-land, free from marshes and miasmatic fever-beds. The forest is not, as is usual in the tropics, so dense as to check or hinder the life of ordinary undergrowth, nor are the sun's rays prevented from reaching the soil and cleansing it from miasmatic poison and bacteria. To avoid taking the disease-germs into one's system, the traveler must abstain from drinking stagnant water, except it first be boiled, and from eating any of the numerous delicious fruits, save only the lemon, lime and orange. The saccharine fruits cannot be indulged in at all, as the bacteria adhere to their surface. To sleep always under shelter, in order to avoid the heavy dews, is a simple yet very sensible precaution. If the tourist follows these dietetic and hygienic rules, he is as safe as in any other country.

Every climate has its own laws, and he who will not obey them must take the consequences. For example, the soldiers of the Federal army perished by thousands in the South, not so much from the effects of the climate as from exposure

and from lack of wholesome food. Again, a man from New
Hampshire, unused to eating watermelons, devours one in
Kansas and dies of dysentery, and his friends say it is the
climate ! The climate of all countries is more uniform
than is commonly supposed ; and the tourist who accepts
these few suggestions on entering the great tropical garden
of Honduras, will there be brought into contact with such a
variety of vegetable and animal life and natural scenery as
cannot fail to benefit him, both in mind and in body. He
receives a new impulse to activity that shows itself in exhil-
arated spirits, always conducive to vigorous respiration and
circulation. Furthermore, the clear soft atmosphere of the
higher altitudes invariably acts as a sedative to catarrhal
affections of the respiratory organs. Inflammatory diseases
of these organs are unknown there. All the organs are
relieved from pressure, and the heart is stimulated to a
more energetic action by the pure, rarefied atmosphere.
Many are the victims of that dread disease, consumption,
in our severe northern climate, who could be relieved and
restored by the wonderful climatic therapeutics wherewith
nature has endowed this mountainous region. Diseases
of the digestive apparatus are in most cases of a bacterial
origin, and caused by indiscretions in food and drink, or by
disregard of hygienic rules, resulting in a disturbance of
the nervous system.

    The problem of acclimatization is easily solved if the
laws of hygiene are obeyed ; no alteration of the normal
condition of the different organs need occur. Precaution
during the rainy season is necessary to protect the system
from the effects of sudden changes. This may be accomplished by the use of light flannel underwear. Before
the rain-fall, an oppressive heat prevails, the whole
body becomes wet with perspiration, and the cool breeze
which usually follows will render one liable to chill unless
the skin is thoroughly rubbed and the clothing changed.

The explanation of this is simple and familiar. Perspiration stops the circulation, and by contracting the blood-vessels, drives to the internal organs the greater part of the blood. Hence arise congestion and inflammatory affections. By a daily sponge-bath, or a bath in the rivers, afterwards rubbing the skin with coarse towels, and by the use of suitable clothing, the skin can be kept in an active and healthy condition, thus averting all danger of this kind. If a chill is felt after a rain or cool breeze, one should at once rub briskly the entire surface of the body. Follow this up with physical exercise, and take, if available, five grains of sulphate of quinine. Reaction will follow, and the healthy relation between the skin and internal organs be re-established. Thus every man has at his own command the remedy for this malady.

Animal food, more especially fat, should be taken sparingly, as it retards digestion. This in turn tends to produce biliousness and other disturbances of the digestive organs, which may develop into more complicated disorders. Hard work of any sort, traveling or other vigorous exercise, should be undertaken only in the early part of the day, in order to avoid the heat at noon and its consequent waste of strength, and to preserve the tone of the physical economy. Sleep, "tired nature's sweet restorer," should be allowed time enough to do its work thoroughly. "Early to bed" is a rule to be guided by here, as well as "early to rise."

Those who live in Honduras maintain that it is the garden spot of the world; while those who visit it bring back reports tinted in glowing colors. It may be said in conclusion that the climate is made up neither of cloud nor of sunshine, but of a happy combination of the two. The elixir of life has not yet been discovered in any of its healing plants; neither, on the other hand, are the entire contents of Pandora's box to be found within its borders. Men may or may not live there to be three-score years and ten. The contributor of this chapter has spent part of his life in Europe, part in the United States, and part in Honduras. His health has never suffered from climatic shock. It is well-known that men travel around the world and live on to a ripe old age. This goes far to show how absurd is the popular notion of the fatality of a climate near the equator, or in fact of any climate within the temperate or torrid zones. A great portion of Honduras enjoys the climate of the temperate zone, and supports in abundance the vegetable and animal life with which we in our own country are familiar.

In order to illustrate the influence of the climate, as above described, upon the conditions and requirements of labor in the lower forest region, and to show how the two climatic seasons govern the choice of time for active work, we subjoin an account of the cutting of mahogany trees. This will afford an insight into the nature of an important and steadily increasing industry. It should be remembered, however, that precisely the same limitations, as to labor during the hours of day-time, do not obtain in any of the mining districts, owing to the greater elevation and lower temperature which are uniformly characteristic of the table-lands.

The mahogany season commences in August, as the wood is not then so apt to split in falling, nor so likely to "check" in seasoning, as when cut earlier in the year. Laborers are divided into gangs of from twenty to fifty men, with a "captain," who acts as superintendent and paymaster. There is also a "hunter," whose duty it is to select beforehand a group of trees, to which he will guide his comrades through the forest. This selection must be made secretly from the top of some tall tree, and requires much shrewdness in order to prevent another gang from discovering the "cutting ground" and appropriating its treasures.

The tree is commonly cut about ten or twelve feet from the ground, a stage being erected for the axe-man engaged in leveling it. Accidents rarely happen. The trunk, from its dimensions, is deemed the most valuable; but for purposes of ornamental work, the limbs or branches are generally preferred, their grain being much closer, and the veins richer and more variegated.

A sufficient number of trees being cut, the preparations for "trucking" begin by the opening of a road to the nearest river. Auxiliary roads and bridges must often be built, requiring great labor and considerable time. In this way the men are kept busily employed until December, when the trees are sawn into logs of various lengths, in order to equalize the loads which the oxen are to draw. A rest is then called until the following April or May, when, the wet season being now on the wane, the trucking begins in earnest. The number of trucks worked is proportioned to the strength of the gang, and the distance generally from six to ten miles.

Take a gang of forty men, capable of working six trucks, each of which requires seven pair of oxen and two drivers, leaving sixteen men to cut food for the cattle and twelve to load. The night hours are chosen for work, in order to avoid the heat of the sun. Loading begins at about midnight, and is accomplished in three hours by means of temporary inclined planes. The drivers now return to the river-side by torch-light (having left there with empty trucks at six the evening before), reaching the chief establishment generally by eleven in the morning. There the logs are marked on each end with the owner's initials, and thrown into the water. During the balance of the day, all the gang are resting, to gather strength for a repetition of the routine above described.

At about the end of May, the incessant rains put a stop to the trucking; the cattle are turned into the pasture, and the trucks housed. Towards the last of June, the rivers swell tremendously, the logs float down stream a distance of, say, two hundred miles, followed by the gangs in pit-pans (a kind of flat-bottomed canoe), to disengage the logs from the branches of overhanging trees, till they are safely lodged in some situation convenient to the mouth of the river.

Each gang then separates its own cutting by the mark on the ends of the logs, and forms them into large rafts, in which state they are brought down to the wharves of the proprietors. Here they are taken out of the water, and undergo a second process of the axe to make the surface smooth. The ends, which frequently get split and rent by the force of the current, are also sawed off, after which the mahogany is ready for shipping.

ROAD MAKING IN THE MOUNTAINS.

CHAPTER IV.

POLITICAL CONDITION.

THE republics of Central America have long been regarded either as nations warring with each other, or as being absorbed in civil dissensions. Their past is full of such incidents, but their present is not. Since the revolution of 1821, under the leadership of Morazan, which resulted in the separation of Guatemala, Honduras, Salvador, Nicaragua and Costa Rica from the mother-country, Spain, these people have been learning self-government.

At first it was a question of mere following, in order to determine who should rule. Like the old feudal lords of Europe, each man of wealth and intelligence had his followers. It is an accepted historical fact that the early English kings were, as a rule, little else than pirates, and that it was not until the reign of Henry VI. that the government took a rational form. The change then brought about was not due so much to the success of the house of Lancaster, as to the revival of learning. Learning in England, prior to the sixteenth century, had been almost wholly confined to the clergy. Men then, however, began to think for themselves. Up to this time the nobility were little better than robbers, the peasantry little better than slaves.

Now, the same conditions existed in Central America in the early and middle part of the present century, excepting that the claims of royalty had been overthrown. The wealth and intelligence of the country was possessed by the few, and they ambitious for power. A united republic was not permitted to stand. The result of the disintegration was five separate ones. Even then, spirited contests arose, of the nature of revolutions, in order to determine who should rule. He who became president was virtually dictator, and the so-called republic an absolute monarchy.

In later years these conditions have been greatly changed. Constitutions have been adopted and their provisions are carefully maintained by those in power. In some of these constitutions it is provided that no president can be his own successor. He may be re-elected, but some other must directly succeed him before he again becomes eligible to the office. Such and like provisions are to-day religiously kept. The growing intelligence of the people demands it. These rules and methods of government are subject to the will of the people. Naturally those in office are not wont to circumscribe their own powers. The people must do that. To-day the true idea of a republic, based upon the principle that they who govern derive all their just powers from the consent of the governed, is gaining ground in Honduras. It may take years to develop fully. That is inevitable. But the influence of this limited democracy is now strongly felt, and the forward strides these republics have made in the past fifteen years are both marked and permanent.

Education is now generally compulsory, and free schools, although primitive in character as yet, are found in every village. The college in Tegucigalpa, Honduras, has a very fair course of studies, and the university at Guatemala City has, in addition to the course for the degree of Bachelor of Arts, a school of law, and one of medicine.

To this university a number of indigent students from each department of the republic are appointed, and the entire expense of their years in college is met by the government. Hence, the advantages of the best institution in the country are not only open to the poor, but are distributed throughout the republic, so that the educational influence becomes general. Latin, English and French are taught in these schools; and there is hardly an officer in a prominent position, but speaks, besides his native language, French or English, or both.

The effect of this more general culture is to put an adequate check upon the exercise of despotism in the government of the people and upon any tendency to anarchy among them. The necessity of good laws and the importance of enforcing them are principles now fully recognized. All shrink from war, and as a consequence these republics are to-day as secure in peace as most nations of the world. The rights of property and of personal security are regarded as sacred, and punishment for violation of law is speedy and severe. Occasionally filibustering bands of adventurers may attempt to overthrow the government, but they are easily put down. Certainly their attempts are unworthy the name of revolution. They are riots, led by lawless men, such as occur at times in the cities of the most civilized and enlightened countries of the world.

Besides the error of confounding the condition of these republics to-day with that of their earlier history, another mistake is made by those who assume that the same kind of civilization and government exists in all the Spanish republics of this continent. Mexico, for instance, has its armed bands of desperadoes, but such organizations have no existence in Central America. The native laborer may be unskilled, ignorant, lazy; but he is always simple, willing, obedient and teachable.

Under more permanent and progressive forms of government, the people are being awakened to the fact that the advancement of their country depends not wholly upon themselves. There are resources of wealth in the mines and in agriculture, to develop which fully will require the concentration of large capital, the importation of machinery, and the employment of skilled workmen who have enjoyed a more special training than their country affords.

The local governments began some ten years ago to encourage the North American to come to the rescue. Concessions were made granting special privileges to companies formed for the purpose of inducing foreign capital and skill to work there. This giving of grants was a natural outgrowth. It was not the purpose of the president of a single republic only. It was rather a common sentiment of the people, which grew out of their own intellectual development. The president granted the concessions and the congress confirmed them ; but it was after all the intelligence of the people of the nation, that really demanded that they should be made. They who owned mines or other properties felt the need of this foreign element to make them valuable. The merchant knew that emigration meant an increase of business for him, the workmen that labor would be more abundant and trade more liberal. Those who governed, therefore, only followed out the wishes of the people in their liberal grants to foreigners.

In this brief survey of the history and development of these republics, the question as to the security of foreigners in their investments there, is already completely answered. The fact to be noted is not so much that the foreigner is seeking concessions, but rather that these Central American countries are seeking him. The political wheel of fortune may place a conservative or a liberal at the head of the nation, but this desire and determination to deal fairly and liberally with foreigners who bring capital and skill into the country have become a settled principle in the national economics.

Unfortunately, it has often happened that adventurers have sought and obtained concessions, and have done nothing under them. About the time of their lapsing, petitions have been sent on praying for an extension of time in which to commence operations, and these petitions have been denied. Then these same adventurers are heard loudly denouncing the government for denying their requests. But while the negotiations were in progress, the republic through its representatives investigated the record of the applicants ; and, if petitions have been denied, it has been plainly because the petitioners proved unable to perform the conditions necessary to render valid the desired concessions. In extended business relations with these countries, the writer has always found them ready to extend every favor possible, when justified in so doing. Their history shows that privileges granted under any concession have been enlarged rather than abridged, by succeeding administrations, in response to fair and upright dealing.

To illustrate these generalizations, a short history of some of the companies represented by the writer as attorney will be in place. The Yuscaran Mining Company was formed upon a concession granted by the Honduras government to Mr. Thomas R. Lombard, of New York, during the administration of the Hon. Marco A. Soto, in May, 1882. Under this, the company was granted six historical mines in the district of Yuscaran (mines which had formerly been abandoned because they had been worked to the waterline), with the *proviso* that work should be begun in a business-like way within two years. Work on one of the mines was to be deemed a full compliance with these terms. In November of 1882, work was commenced by sinking a shaft on one of these, the Quemazones, and the vein was touched early in the following spring.

The Hon. Marco A. Soto resigned the presidency of Honduras in September, 1883, and an election was called in accordance with the provisions of the constitution. At the election, General Luis Bogran was elected constitutional president, by the vote of the people. General Bogran shortly afterward wrote to Mr. Thomas R. Lombard, the General Manager of the company aforesaid, promising to do everything in his power to foster foreign investments in his country, whether made under concessions granted by Mr. Soto, or under those of any former president. Prior to the resignation of Mr. Soto, liberal concessions had been granted to the Central American Syndicate Company by the Honduras government, and up to the time of the election of General Bogran nothing had been done under them. This assurance of President Bogran gave confidence to investors, and soon the Paraiso Reduction Company was formed, under a grant giving it the sole right to establish custom reduction works in the Mineral de Yuscaran, together with other valuable franchises.

During the three succeeding years, other companies have been organized, and have commenced active operations, there being about a thousand native laborers employed at present at the works of the several organizations. During this time President Bogran has been found always ready to give the representatives of these companies an audience and to aid them in the active prosecution of their interests. The greatest favor asked for was the building of a wagon-road from the Port of San Lorenzo on the Pacific coast to Yuscaran, *via* Tegucigalpa, a distance of about one hundred and twenty-five miles. Pack-mules over the mountain-paths had carried all the machinery up from the coast, excepting pieces weighing over two hundred pounds, which had been tied to poles and carried up by men. Such a road would prove, of course, useful to the government, but it was absolutely indispensable to the success of these enterprises. General Bogran, in the liberal progressive spirit that has characterized his policy, recognized this need, and through his prompt action the government undertook the work. The building of this road occupied eleven months and cost Honduras nearly one hundred thousand dollars.

All the concessions made during the administration of Mr. Soto have been ratified and confirmed during the present administration. New and valuable ones have lately been granted to the Central American Syndicate. Marked courtesy has been shown to all interested parties visiting Honduras, by President Bogran and by other officials of government, as well as by private citizens of the republic.

In visiting these countries, the writer has made careful inquiries as to the several republics, in regard to the general security of investments. It was found that by them all the self-same favor and protection is uniformly extended.

A prominent citizen of Balize, in referring to certain losses of property suffered in Honduras about fifteen years ago by subjects of Great Britain, declares that these losses occurred at the time of a revolution, and that upon the presentation of the claims to the government they were promptly made good. An Englishman who has a coffee plantation in the republic of Salvador where he employs three hundred and fifty natives, and who has been in business in this and in the other republics for over twenty years, gives his personal testimony that the policy uniformly is to protect foreigners in their rights, and that new presidents are zealous in even adding to the rights and privileges granted by their predecessors.

Americans, Englishmen and Germans of long experience in business dealings in Central America, bear unanimous witness to the fact that these republics welcome and protect all foreigners working under government grants.

The Hon. Henry C. Hall, United States Minister to Central America, a man of long diplomatic experience in these Spanish republics, avers that he can recall no case where an appeal to the United States government was needed to secure to any of its citizens the full enjoyment of the most liberal construction of the rights and powers granted them in concessions made at any time.

Since the recording of the concessions to the Syndicate in the State Department at Washington, the Hon. Secretary of State has asked the United States Consuls in Honduras to make special reports on the condition of these and other business concerns there, in which citizens of this country are interested. This action of the State Department is, in itself, an assurance that our own government will take a more active interest hereafter in the protection of its citizens in their investments abroad under franchises received from other nations. In fact a United States Consulate has been created at Yuscaran, simply because of the large operations carried on in that particular district of Honduras by United States citizens.

Since April, 1881, the writer has represented in a legal way a number of New York companies engaged in business in Honduras, and thrice during that time has visited the country. During these six years the representatives of these companies have received complete protection and many wise suggestions from the executives and officials of the republic.

The prospect of the future must be judged not merely by the records of the past, but by these records coupled with the progressive national sentiment of the people. This sentiment has been seen to be most favorable to the foreign element engaged in business in this locality; and the respect for law and justice, which is born of intelligence and grows with it, has become in itself a complete guaranty of the security of such investments.

TUNNEL—SAN ANTONIO MINES.

CHAPTER V.

AGRICULTURAL RESOURCES AND NATURAL PRODUCTS.

FEW countries of the world possess natural advantages of climate and soil equal to those of Honduras. Comparatively little labor is needed to produce any of the crops of the torrid or temperate zones. The harvest which rewards industrious cultivation will yield a rich income to the agriculturist.

For example, in the level lands of the coast region, sugar, indigo, and cotton grow abundantly in response to the most moderate encouragement. Sugar does not require to be replanted as in other countries, but will bear crop after crop for many years. Again, there is an important trade in tropical fruits, chiefly in bananas and cocoanuts, large quantities of which are shipped every year to the United States.

Then there are immense resources needing little or no expenditure of capital or of labor upon the soil. Various kinds of valuable woods flourish in the forests; mahogany, cedar, log-wood, the india-rubber tree, and several varieties of hard wood similar to rose-wood and ebony, suitable for the purposes of the builder and the cabinet-maker. Fibrous plants are found in some districts in the greatest profusion, affording a wealth of raw material for the manufacture of ropes and cordage.

In the interior of the country the wide ranges offer unsurpassed facilities for the grazing of cattle, an industry long appreciated by the inhabitants, and already developed to a considerable degree.

Cotton.—This staple was successfully cultivated in Honduras during the period of civil strife in the United States, and the product was pronounced a very superior article. The conditions of the country are favorable to its growth. There is little doubt that the raising of cotton in Honduras could, if rightly undertaken, be made very profitable. It was tried on a small scale by an American planter who went to that country shortly after the close of the war of the Rebellion, and settled in the town of San Pedro Sula. Several acres of cotton were grown by him from seed brought from Georgia, of the variety known as the Sea Island cotton. The stalk was about eight feet high, and the bush averaged fifteen feet in circumference. In certain seasons it is covered at the same time with open boles, immature ones, and blossoms.

This gentleman gathered, from this field, cotton to the amount of five hundred pounds to the acre, giving him four pickings a year ; and owing to the absence of frost in that section, he never was troubled with yellow cotton. The chief difficulty he had to contend with was getting pickers for it, the people not understanding how to do the work properly. If, however, some one having sufficient capital were to plant a large plantation and import southern negroes from the United States, who understood cultivating and picking the article, and were then to put up his own gins and presses, a fine business could be opened that would pay large profits on the investment. This particular cotton had been planted twelve years, and there appeared no reason why it should not yield for several years longer, although it was decreasing slightly in its yield, the growth going mostly into the trunks and branches. There is, in some of the gardens of the country, a native

AGRICULTURAL RESOURCES.   45

cotton which grows on a long vine. There is said to be a
pink variety of this, that is to say, a kind giving a pink fibre,
instead of the white one so familiar in the fields, the press-
rooms and warehouses, and the marts of trade.

Sugar-Cane.—The principal product of the department of Comayagua to-day is sugar-cane, which is now
cultivated to meet the limited demand of the local markets.
It is indigenous to the soil, and very luxuriant. In Olancho,
one witness speaks of sugar-cane growing without care upon
the plains during a period of thirty years, while a political
disturbance was in progress, and declares that this valuable
plantation failed only for the want of men to gather in the
crops.

In nearly all departments of Honduras, in fact, the
climate and soil are both eminently adapted to the growth
of sugar-cane. At present this industry is confined in most
quarters to the small growers who cultivate it for the purpose of producing *aguadiente*, a native rum. Under the laws
of Honduras the latter article is a government monopoly.
Any person desirous of making the same can do so by
getting permission from the government, and agreeing to
sell all of the manufactured article to the government,
which in turn retails it, through agents established at each
town, to the people. It is sold by the bottle at about six
shillings, Central American money.

The sugar cane itself is of a superior quality, producing large white crystals when refined with care, and only
needs replanting once in ten or twelve years, according to
the locality where it is grown. Almost every farmer
throughout the country raises a small patch of cane for the
purpose of feeding his stock with it. A few years ago a
gentleman from Balize obtained a concession from the government for a large tract of land in the valley of the
Sula, proposing to raise sugar-cane and to establish a

refinery there. He went to England, expecting to secure
the necessary capital for the purchase of machinery, etc.,
but was unsuccessful. There is little doubt that this country could be made one of the largest producing districts for
sugar in the world, provided that capital and intelligence
were directed to that end.

Coffee.—Coffee of excellent quality flourishes freely in
Honduras, although it has never been adopted as an article
of general production, not even to the extent of supplying
the people of the country. The Department of Gracias, for
instance, has some coffee gardens; but while the bushes
are heavily laden with berries in their season, proper care
is not taken of the crops, chiefly for lack of workmen to
attend to them.

As a record of actual and recent results in this branch
of agricultural industry, it may be mentioned that out of
one million sacks of coffee exported during the year 1885
from all Central America, Honduras furnished twenty
thousand sacks, without any special exertion.

There are, in fact, people of almost all nationalities,
including principally Germans and Italians, at present
engaged in planting, raising and gathering coffee, cocoa, etc.
Some of the coffee raised is shipped, along with other products of the country, to American ports. But the greater
part of the export, as well as of the import trade, is carried on
with England and Germany, as has been the case for many
years. The plains of Comayagua are admirably adapted to
the cultivation of coffee and kindred staples. There is
every reason for believing that coffee of equally good quality with that of Costa Rica may be produced in Honduras,
which possesses every requisite variety of soil and climate.

Fruits.—For a brief survey of the traffic in tropical
fruits, one of the chief industries of Honduras, the depart-

ment of Colon may be taken as a representative district; fruit culture for exportation being extensively carried on through all the region watered and drained by the Guayape river and its tributaries.

Here the simple native willingly assists, for the smallest compensation, in gathering for distant markets the abundant treasures of the tropical orchard-fields. Bananas, custard-apples, rose-apples, grapes, plums, limes, lemons, oranges, pomegranates, citrons, melons, pawpaws, mangoes, *guavas*, *maranones*, *aguacates*, *achiote* and cocoa ; all these contribute to the rich store of native delicacies from which the dealer makes selections for his foreign customers.

The banana, in fact, grows wild in the country. Five kinds of cocoa are found, each with its specific use and value. Of the *sapote* there are four varieties; of the *aguacate* three. Of the *guava* fruit, the favorite is known as the "arrallan," growing on the savannas or plains ; this kind is preferred to the common as well as to the Peruvian variety. A very good quality of olive thrives in this country, besides a similar fruit, *negritos*, commonly called "the olive." Besides these fruits, the vegetable products of the soil include tobacco, indigo, sassafras, Peruvian bark, vanilla, ipecac, pimento, ginger, pepper and sarsaparilla. For provisions, the kidney-bean, potato, rice, wheat, corn, the yam, which grows here to an immense size, and the plantain, next to the maize the principal reliance of the people of the tropics as an article of food. The yield of the plantain is simply enormous, the product of a single acre, according to Humboldt, exceeding the crop of a hundred acres of wheat, or of forty acres of potatoes.

Yoro is one of the north coast states of Honduras, and second in size to Olancho. Its principal wealth is in its agricultural resources. Especially abundant here are the banana and cocoanut. The great plantations of Puerto Sal are said to yield over $100,000 worth annually of cocoa-nut oil alone. The principal cocoa-farm is eighteen miles wide, and counts over 22,000 productive trees. About the same number are growing on the Ysopo farm.

Elaboration by machinery produces from the cocoa, in addition to the oil which is so valuable an article of commerce, four different kinds of tow. After the oil is extracted from the pulp, the residue is utilized as a desirable food for cattle.

Another large industry is that engaged in the production and exportation of sarsaparilla. This commodity, which commands a ready and extensive sale in nearly all countries of the world, is readily obtainable in Yoro, as elsewhere in Honduras, and from this department valuable consignments are made every year to the wholesale dealers in the United States and Europe.

At the present time, a railroad is said to be projected by Western capitalists, having for its object the opening up for trade of the whole coast line of Yoro, between Truxillo and Puerto Cortez. The principal traffic contemplated is in fruits, chiefly bananas, cocoanuts, and pineapples, for the United States.

Precious Woods.—At the present time, the many varieties of precious woods constitute an important item in the exports of Honduras. Those best known are the cedar, the mahogany, and the rose-wood. The mahogany grows in nearly all parts of the country, in the valleys of the various streams. It is, however, most abundant upon the low ground bordering the rivers that flow into the Bay of Honduras, where it also attains its greatest size and beauty.

In the valleys of Comayagua, there is any quantity of mahogany, *lignum vitæ* and cedar. The finer qualities found along the banks of the Sulaco river, are so far distant from the coast as to have little or no commercial value. Those varieties which grow in the mountainous districts are the tallest and of the greatest diameter, and thrive alike in all temperatures. The pine and oak are abundant on the hills. The department of Olancho has even greater variety to show in forest products. There is one wood called *ronron*, held to be unrivalled for the manufacture of handsome and durable furniture. In Juticalpa, the capital city of Olancho, the resident artisans have utilized this superior material in the construction of household furniture, such as sofas, chairs, bureaus, and cabinets. By so doing they are enabled to compete with foreign dealers in these necessary articles.

This district furnishes also large quantities of walnut, black-wood, tamarind, mulberry, log-wood, and other dye-woods; several varieties of oak, including the evergreen oak, the sturdy San Juan, and Guanacaste; fig tree, *ceibas, granadillo, guayacan, teguaje, alazar, espino verde, masicaran, chihipaete, coyote* and *guano*, the last named being a sort of native American palm. All these flourish in abundance, in addition to the forests of mahogany and cedar, which almost everywhere prevail.

Some of the above-mentioned trees produce food for both man and beast; some yield a species of cotton well adapted for household uses; and others are exceptionally rich in resins, gums, and oils for various purposes. Among the latter may be specified gum *copal, balsam, talascan* and *guapinal*. The *caoutchouc* is found in sufficient quantities to repay constant attention, its product being used for making water-proof cloth. There are also endless tracts of pine forest, in part situated conveniently near the sea, and along the banks of rivers navigable by steamboats, making them very easy of access. In the same neighborhood, from the coast to the confluence of the Guayape and the Guayambre rivers, are great forests of balsams, limas, cedars, and mahogany. Innumerable varieties of dye-woods shade the wide valley of Juticalpa, crossed by the Guayape.

Whoever shall be enterprising and persevering enough to open up the districts traversed by this river, bringing its magnificent timber-lands into tribute to his business energy, will reap a rich harvest in return for his exertions.

Vegetable Fibres.—Of the different kinds of fibrous plants so prolific throughout the country, those deserving of special notice are the *maguey* or *agave* (commonly called *mescal*), the *pita* plant, and the *junco* or Panama straw, flourishing chiefly in the department of Comayagua.

The *mescal* leaf is about three feet six inches long, white and strong in fibre. The *pita* is finer, of a silky texture and lustre, about four feet long, suitable for mixing with silk fabrics, as proved by experiment. This fibre is exceedingly tough, and is capable, under proper treatment, of almost endless subdivision into the finest threads. The *junco* is the well-known water-plant of which Panama hats are made.

The following careful description of one of these fibres will prove interesting reading :

Pita.—The *pita* plant differs from an aloe proper (the European aloe belongs to the lily family), and also from the cactus, with both of which it has too frequently been confounded. Mr. Richard Whiting, in a letter to "The New York World," some three years ago, speaks of the planters of Mauritius cultivating it for its valuable fibre, and names it a species of aloe (*aloe Mexicana*); and Dr. Trowbridge, United States Consul at Vera Cruz, in 1880, is reported as saying, in speaking of a variety of this plant, if not actually the same species : "There is a species of cactus here commonly called *pita* (I do not know its botanical name), some of the fibres of which are sixteen feet long. It is strong and silky, and capable of being drawn into threads, from which gossamer webs might be woven. In fact a few months ago a Vera Cruzan sent some of the fibre to England, and had a few handkerchiefs made, which were extremely beautiful, and appeared more like silver tissue than linen, and were quite strong." Both Mr. Whiting and Dr. Trowbridge were mistaken as to the botanical name, although correct in what they say in reference to its adaptability.

Pita is most generally known as the American aloe or agave plant (*agave Americana*). It belongs to the amaryllis family, and it has been put to a great variety of uses in southern Mexico and the several republics of Central America. The dried flower-stems have been extensively utilized to make thatched roofs for tropical houses, the strength of the fibre giving to such roofing wonderful durability. The sap of the leaves of one species of the American aloe (a coarser species than the one which is the subject of this article), becomes, when fermented, the well-known Mexican drink *pulque*, and when distilled, the pleasant but deceptive *vino mescal*.

The fibre, which is the most valuable and wonderful quality of the *pita* plant, has been extensively used by the natives for hammocks, cordage and ropes. It extends the entire length of the leaf, and the natives extract it by first pounding the leaf on a rock, next exposing it to the rays of the sun (whereby the bark of the leaf becomes crumbled), and then giving it a second pounding followed by a combing, which produces a clean fibre. This process is necessarily slow and expensive, which accounts for the fact that the use made of the fibre has been almost entirely confined to the tropical countries producing it.

The *pita* plant of Central America seems to yield a finer fibre than that of Mexico; however, there is a marked difference in this particular in localities on the same parallel. In the lowlands of Honduras and Nicaragua, where *pita* grows most luxuriantly, the leaf is straight, varying from three to four inches in width, having no middle stalk, and from a few feet to eighteen feet in length. In its growth here it monopolizes the soil, taking exclusive possession, excepting, of course, the space occupied by the tropical forest trees.

From the fact that this plant has, as yet, received no systematic cultivation or cutting, it is impossible to determine the exact annual yield; but, from the best available sources, this yield (by cutting the leaves three or four times a year, so as to bring up an average of six feet in length), will be from three to five tons of clean fibre per acre. The territory occupied by this plant is exceedingly extensive. Along some of the water-courses, and extending back from them, single tracts can be found containing a thousand acres. The crude fibre is equal in value to manilla hemp, when applied to light uses; but in fineness, strength and durability, it is far superior. The ultimate fibre is even finer than that of the threads of silk spun by the silk-worm.† Experiments have been made of weaving this fibre, when flossed, with cotton, wool, or silk; and it has been found that this can be done advantageously with any of them. As the *pita* fibre possesses a silky gloss of its own, it has been thought by manufacturers that it would be found valuable to mix with silk, especially in the manufacture of heavy curtain fabrics where weight, strength, durability and finish are required.

Companies have been organized for the purpose of bringing this fibre into the market for general use, and considerable sums of money have been invested in machinery for separating the fibre; but, at present writing, complete satisfaction with any of these machines has not been reached in the tests made; and a wide field seems here to open itself to the inventor. Years of patient work of inventive

---

† The writer of this was shown the two under a powerful microscope at Lyons, France, and heard many exclamations of surprise on the part of manufacturers at this unexpected result, and at the further fact that the pita fibre did not lose its strength when reduced to the fine floss state.

AGRICULTURAL RESOURCES.   53

brains were required before proper machinery was perfected to produce *remi* fibre from the stalk, and some waiting may still be necessary before a practical machine shall be completed capable of reducing the green leaf of the *pita* plant to a marketable product. When that is done, a new and valuable fibre will have daily quotation in our market-reports, and manufacturers will find for it a hundred uses, not only because of its wide adaptability, but because it is found, by actual tests, to be the strongest vegetable fibre known.

Cattle.—Cattle, hides and deer-skins form the chief exports of the department of Olancho, relatively, and for this reason, the richest of all the departments of Honduras in actual available property. It exports also considerable quantities of sarsaparilla, tobacco and bullion; but the gold-washings themselves are secondary in value to the cattle-herds as a source of wealth to this region. Vast herds wander over the spacious cattle-runs of Olancho, finding, in the wide savannas and open forests, ample pasturage and congenial roving-grounds. The ox grows to a size above what in our own country is regarded as a good average, and is of remarkable beauty and strength of form, short, compact and powerful. The cows do not as a rule yield a large quantity of milk, but its quality is good. These animals are successfully raised in various portions of Honduras, and constitute an important share in the property of the people. Large numbers of oxen, broken to the yoke, are supplied to the mahogany-works along the north coast.

Exports of cattle are made chiefly to the neighboring states of Central America and to Cuba. Guatemala and San Salvador consume a full proportion, the latter country, in common with Balize, drawing nearly all its entire supply of cattle from Honduras. Numbers are sent also every year to Panama. Hides are destined soon to become a still more important article of export, owing to the improved methods of transportation now being introduced into the country.

The general character of the country, abounding throughout its entire area in natural and unfailing meadows, affords immense facilities for cattle raising, and is eminently favorable for the increase of this kind of property to an indefinite extent.

Opals.—The omission of one minor point of peculiar interest would be deemed inexcusable. For this reason attention is called, at the close of this chapter of miscellany, to the opals of Honduras, celebrated by the gem-loving cavaliers of Spain in both song and story.

The opal mines of Honduras are situated in the department of Gracias. They have been worked, principally by foreigners, for a number of years. The fire-opal of Honduras takes equal rank with that of Hungary. Some specimens are very beautiful, although as a rule the ordinary or milk-opal is a very insignificant-looking stone.

At an exhibition of the products of Honduras recently given in San Francisco, a collection of opals belonging to ex-president Soto excited great interest. Some of these were very large and beautiful, and carved in different shapes. One represented a tortoise, another a *scarabæus*. A larger specimen, said to be the largest precious opal ever found, was a miniature castle situated on a cliff,—the whole being about five inches high and three inches square at the base. A small fire-opal found by Mr. Peacock in his mines a few years ago, was so very beautiful that, although not larger than an ordinary white bean, it brought $350 in the London market.

It seems a pity that these gems should have been so long excluded from the American jewelry-market solely because of the absurd superstition regarding them. This familiar superstition is of comparatively recent origin, having

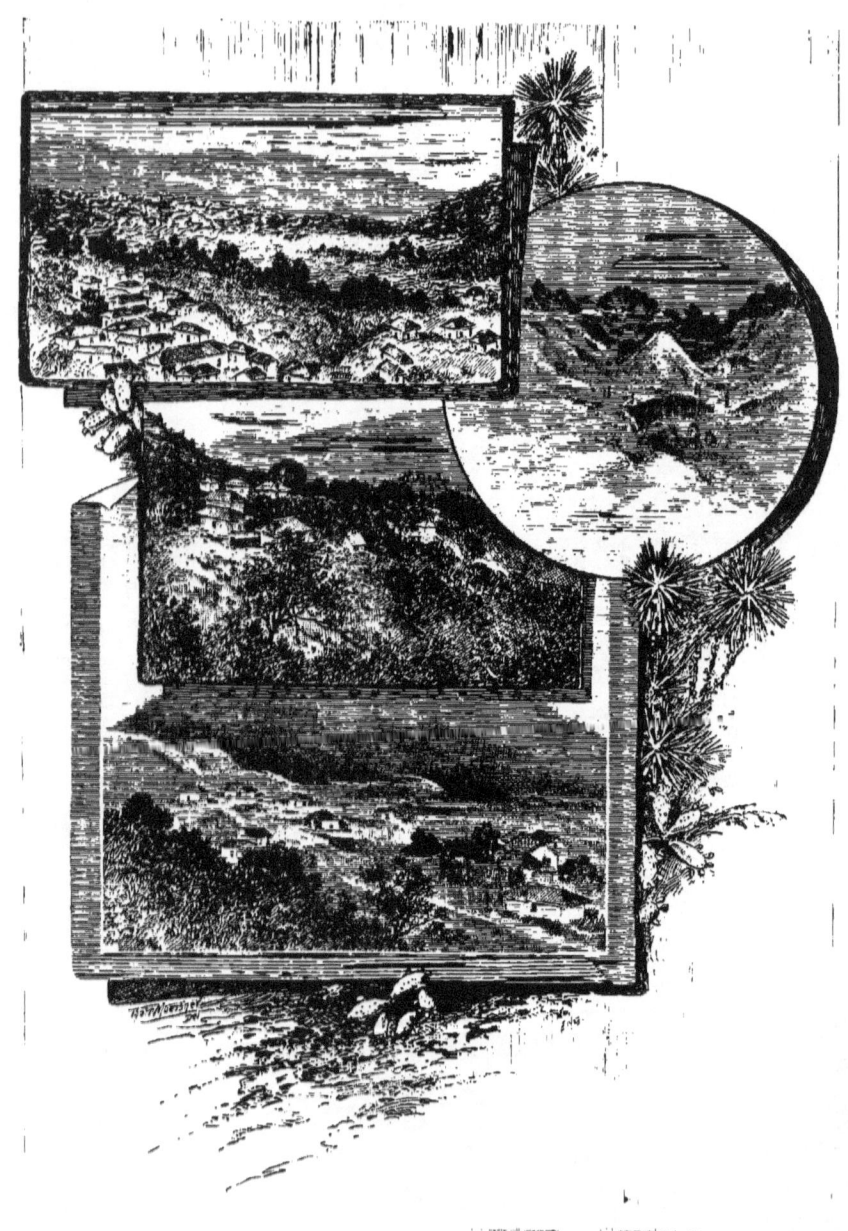

SAN ANTONIO VILLAGE.

AGRICULTURAL RESOURCES.   55

received considerable impulse from the prominence given to
it in a novel by Sir Walter Scott. Afterwards it was
fostered by Parisian jewelers at a time when the scarcity
of the stone made it impossible for them to meet the
demand, their object being to turn the demand away from
this gem to others more available.

The ancients held this stone in very high esteem. It
was long believed to be the luckiest of all the jewels, as witness the following lines :

> " Gray years ago a man lived in the East,
> Who did possess a ring of worth immense
> From a beloved hand. Opal the stone,
> Which flashed a hundred bright and beauteous hues,
> And had the secret power to make beloved,
> Of God and man, the one
> Who wore it in this faith and confidence."

CHAPTER VI.

THE MINING INDUSTRY.—ITS PAST HISTORY.

THE mining industry of Honduras may be said to have begun with the discovery of the country by Columbus in 1502. The attention of the great navigator's followers was drawn immediately to the mineral resources of the country. Their cupidity was strongly aroused at sight of the gold ornaments worn by the Indians who traded with them on the Mosquito coast, now known as the department of Colon.

As a result of that first expedition, the conquest of Honduras was determined upon. Two years later the plan was put into execution. Truxillo, Puerto Cortez, and Omoa were the points first settled, and from these the earliest expeditions were made into the unexplored interior. Next followed the establishment of trading-posts, the *nuclei* of present towns and cities.

The district now known as the department of Olancho, most conveniently entered from the port of Truxillo, seems to have been the Mecca of the Spanish adventurer in Central America. All its rivers and streams flowed over the precious gold-bearing sands that promised untold wealth to the Castilian.

It is commonly believed that the only kind of mining done during the first half-century of the Spanish occupation was that designated *placer*. The virgin streams of the north coast offered the most remunerative fields to the gold-hunter, involving as they did no special outlay of skill, labor, or capital.

The processes of extraction and separation used by the early miners were very simple, being in fact on a level with the intelligence of the conquered Indians, who were forced to toil for the enrichment of their hard and pitiless taskmasters. A shallow wooden dish, not unlike the chopping-bowl of domestic use, was the only requisite appliance. Into this was put fifteen or twenty pounds of river-sand, and water having been added, a centrifugal motion was imparted to this mixture by twirling the bowl horizontally in the hands. The difference in specific gravity of the component parts, under the influence of this centrifugal motion, caused the heavier grains to settle towards the bottom. Gold, being the heaviest, formed the substratum, and the outer component materials were then washed away. To the early Spanish settler in Honduras the problem of mining was one to be solved by simple hand-labor.

As to the amount of hand-labor available at that period in the history of the new country, a glance may be taken at some of the records still extant. The priestly chroniclers of those expeditions of conquest and occupation people Honduras with hordes of fierce savages, to subjugate whom required the waging of an incessant warfare.

The tales of wholesale slaughter contained in these records recall the accounts of massacres in the days when the historic nations of southern Europe were struggling for the mastery of the world. Such exaggerated estimates as to the population of the country are scarcely in accord with the complaints, continually repeated by these same chroniclers, of the great difficulty of obtaining laborers to work the mines. But a few years were required, it seems, to subdue the native tribes and to bring them under the influence of the Roman Church. And yet from the outset, till the last vestige of Spanish authority disappeared from the country, the scarcity of labor was the chief obstacle encountered.

In view of this important fact, the old Spanish estimate of the population is hardly to be accepted. It plainly grew out of an overweening desire to magnify the deeds of valor performed by the Spanish soldiers. No excuse is to be offered for the cruelties practiced upon a weak and defenceless people. Much, however, might be said in admiration of that mere handful of determined men who courageously met and skillfully overcame the opposing forces of savage tribes, whose numerical preponderance alone seems sufficient to have annihilated the little band of invaders.

For instance, in a battle fought on the plains of Comayagua, the chief of one native tribe is said to have mustered thirty thousand warriors. This, according to the usual method of estimate, would give to that section of the country a population of at least one hundred thousand, a greater number than could have found sustenance there. Such an estimate, if proportionately continued, gives to the limited area of Honduras a population, at that time, of some three millions, or about ten times what it is thought to be at present. Peopled to that extent, Honduras could not have been the country it was pictured, a succession of dreary plains, wide savannas, and immense forests, ranged by dangerous wild beasts.

The habits of a migratory people lead them to follow the water-courses of a country, for reasons obvious yet different from those which gave to the invaders of Honduras their prime motive. By the attraction of the gold-bearing sands of the rivers, the Spaniards were held to those localities; while the fish of the streams and the game which frequented the river-banks afforded sustenance to the aboriginal. Probably the Spaniard, following the courses of the streams, came in contact with all the Indian dwellers of the country, and made a false estimate of the population of the whole district by allotting to interior and uninhabited portions a proportionate number. Furthermore, it is not to be presumed that where the Spaniards were opposed by a considerable number, the opposition was purely local.

Difficulties between neighboring tribes are often laid aside in order to combine against a common foe. While it

is possible that thirty thousand Indians were gathered on
the plains of Comayagua to give battle to the Spaniards,
it is more than probable that this force was recruited from
all the tribes of the surrounding country. The region in
question is centrally located, and well adapted for the assem-
bling of a large force with intent to make a last determined
stand against the encroachments of an invader. For more
than fifty years, indeed, the tribes held under the sway of the
Spanish sword had suffered untold miseries, and a consolida-
tion of the fighting forces of the natives for one mighty resist-
ance to the terrible tyranny grinding them down into bitterest
servitude, is the natural explanation of the existence of such
an army—if the statement as to its existence is to be received.

Subsequently to the battle of Comayagua, all vigorous
attempts to oppose the rule of Spain seem to have been
abandoned. The unhappy native, ceasing to be lord of the
domain, becomes no better than a beast of burden. Certain
old manuscripts, which have survived the edicts of the
Spanish crown, recite acts of barbarism on the part of the
invading bands too horrible almost for belief. If we are to
take the statements of these eye-witnesses, the story of the
conquest of Honduras reads more like a series of cold-
blooded butcheries than a record of the valorous exploits of
military heroes. So disgraceful to Spanish arms were these
recitals, that the publication of them was interdicted, and
care was taken to put before the people an account which
reflected honor and credit upon both Church and State.

One of the earliest discouragements to the mining
industry was a revolt of the peons employed in the gold
washings. Unable to endure any longer the increasing
hardships of their lot, and aware of the futility of a resort
to arms, the Indian slaves fled to the mountains, leaving
their masters behind them in a helpless condition without
even the assistance necessary in tilling the ground. The
lesson of such a predicament, involving the danger of star-

vation to the Spaniards, was naturally not lost upon them. So soon as by conciliatory measures the natives could be induced to return, the policy was adopted of actively promoting agricultural industries. That this policy was gladly fostered by the Church may reasonably be supposed. In the year 1511, a council to provide for the better government of the Spanish colonies was created under the title of "The Council of the Indies." This Council passed a decree, among others, that a fifth part of the product of the mining industry should be paid to the King of Spain ; while to the Church there should be paid a royalty consisting of a tithe of the agricultural product.

Taking into consideration the indubitable fact that the Church rapidly acquired great wealth in Honduras and that its revenue was derived solely from the tillage of the soil, it is beyond question that labor was largely diverted to agriculture, even to the detriment of "the King's fifth." No other reason so plausible can be adduced for the prevailing distaste for work in the mines, a feeling apparently amounting to a strong superstition, and continuing until after the independence in 1821, when the whilom slave became a miner on his own account.

The best evidence of the scarcity of laborers for this important industry as conducted during the period of the Spanish rule, is to be found in the numerous petitions to the crown, still extant, wherein the mines are shown to be capable of greater production, unattainable for lack of men to work them. Another drawback causing frequent complaint was the drafting of miners in preference to plantation-laborers, into military service on the north coast. The buccaneers who infested the Spanish main, lying in wait for Spanish treasure-ships, often descended upon the coast. Owing to the constant fear of these robber-raids a large force of soldiers was kept in requisition to protect the principal points on the coast. Still another cause for complaint was found in the discrimination made by the authorities in favor of Salvador, in this same matter, to the prejudice of the mines of Honduras. Miners were drafted and sent to the former country, probably because of the proximity of Honduras to Guatemala, the seat of vice-royalty at that time.

The discovery of silver was made some fifty-seven years after the first Spanish invasion, but no attempt was made to open up the mines. The work of conquest was in hand, and the gold-bearing streams and surface-veins offered a readier and more profitable venture. About the beginning of the seventeenth century the opening of the silver veins was undertaken. The mountains around what is now Santa Lucia were the scene of these primary operations. The adjoining districts, or *minerals*, of Santa Lucia and San Juan de Cantarranos (which latter a few years ago was subdivided by the creation of the *mineral* of San Juancito) are designated as the first regular mining-camps established in Honduras.

In these mining districts are to be found to-day evidences of the methods of mining employed by the workmen of those early days. These evidences indicate that such natural difficulties as were successfully removed were overcome by mere brute force. No scientific knowledge whatever appears to have been applied to the task; in fact the degree of intelligence which directed these pioneer efforts could not have been great. From the time when miners were accustomed to split rocks by heating and then suddenly cooling portions of the surface, until a very recent date, Honduras has been as far behind Mexico, in the improvement of its mining processes, as Mexico has been behind the United States in this respect.

In forming an estimate of the mineral possibilities of Honduras, a comparison of that country with Mexico is naturally suggested. But beyond the fact that a common language is employed in the two countries, little is found upon investigation to warrant a comparison.

Prescott shows that when the Spaniard first appeared upon this continent, Mexico was already a well-established empire, teeming with a great population, intelligent and skilled in the arts fostered by an era of prosperity and peace. Honduras, on the other hand, was the home of nomadic tribes subsisting by the chase. Mexico was already rich. Thither the Spanish chieftain, seeking his fortune in America, directed his steps. There he established a close relationship with the old world, and enjoyed the advantages derived therefrom; while Honduras fell to the lot of the rough and turbulent soldier. It was a mutiny of the men under leadership of a headstrong lieutenant that led the immortal Cortez to make his celebrated march from the city of Mexico, through trackless and at times hostile regions, in order to arrest and punish the rebels,—a march remaining to this day without a parallel in military history.

Mexico, with its vast population, providing an abundance of labor, and with evidences of wealth on every side, claimed the immediate notice of the mother-country, Spain. It gained as a result the fullest development possible under the existing conditions of the mining industry. Enjoying this close intimacy with Spain, it is no wonder that Mexico took such forward strides. Honduras was able to communicate with its "protector," making known its wealth and its wants, only with difficulty and indirectly, *i. e.*, through Guatemala, and thence to Europe by way of Cape Horn.

While it is probable that no country, in proportion to its area, contains so much mineral wealth as Honduras, yet like many another colony it was for a long time over-shadowed by more populous neighbors, continuing to contribute its quota to the wealth of Spain. And this it is known to have done without that encouragement which unquestionably, had it been given, would have brought this region into the front rank of Spain's most valuable possessions.

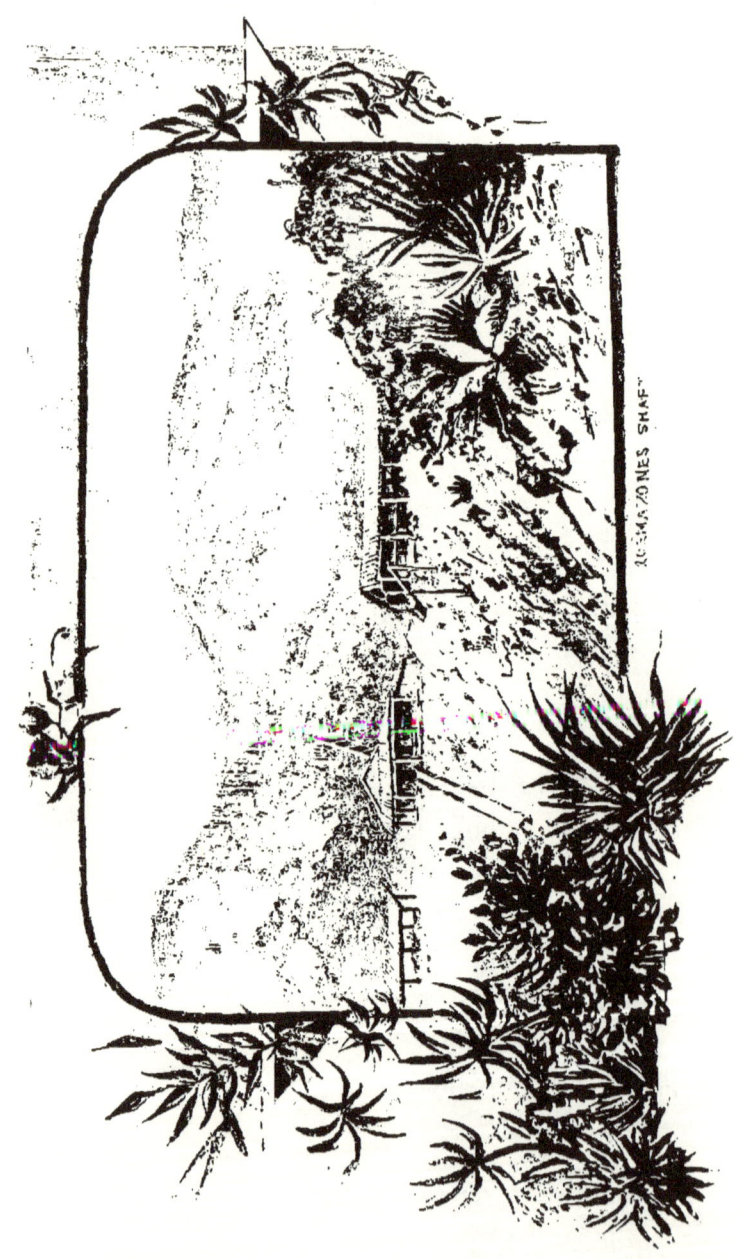
"LOS AZONES SHAFT"

CHAPTER VII.

THE MINING INDUSTRY.—CONTEMPORARY HISTORY.

IN the development of the mining industry, there exists no evidence of great undertakings. Wherever the signs of extensive operations appear, they are surrounded by those natural conditions which would make such operations practicable with the simplest of appliances in the hands of the least intelligent of workmen.

Whenever a state of affairs was reached requiring a system of artificial ventilation or drainage for the sake of further progress, abandonment was the result. Whether such abandonment was encouraged by the fact that there was an abundance of other veins to be worked, or whether it resulted from a lack of knowledge of the proper principles of mining, it is difficult at this time to determine. The fact remains that abandonments occurred when these obstructions were met with, apparently without regard to the value of the property.

This may be observed, for instance, in the case of the Corpus Christi mine in the department of Choluteca. This mine is said to have produced fabulous sums of gold; but the inability to drain off water caused a suspension of work, and to-day the property remains in possession of that obdurate foe to Honduras mining. Native mining in Honduras at the present day is a fair index to that of the past. Where, for example, a ready means of artificial ventilation presents itself, the water-line limits active operation. Again, in the absence of such means of ventilation, a still narrower limit results, arising from the difference between the temperature

64   THE NEW HONDURAS.

within the shaft and that at the mouth of the mine, a difference controlling the flow of air-currents.

How serious a problem this matter of water was to the miner of Honduras in former times, as well as at the present day, may be understood from the following statement.

Ore and water are removed from the mines in leather bags carried by men. In shaft mining, the interior is reached by means of notched timbers, about ten feet long, used as ladders. The shaft consists of a series of sections, or *posos*, so that the whole resembles more than anything else a flight of stairs intended for the use of the giants of ancient fable. In order properly to state the water problem, it is assumed for the sake of illustration that a certain mine has a depth of say one hundred and fifty feet, and that it contains one hundred thousand gallons of water. To reach that depth, fifteen notched poles must be climbed. Hence the force of men employed in raising the water is limited to fifteen, part of these ascending while the others descend the shaft. The maximum of round trips daily made by these men would not exceed forty for each, and the quantity of water raised by each man per trip would not be greater than fifteen gallons or one hundred and twenty-four pounds. Thus nearly twelve days would be consumed by this force of men (the greatest number capable of being employed in such a shaft), in raising the one hundred thousand gallons. Whereas, in the State of Pennsylvania, over three million gallons have been pumped and lifted by machinery in twenty-four hours.

From the foregoing it may be seen that a small amount of water, at an insignificant depth, means the loss of the mine, at least under the disadvantages of the mode of operation that has just been described. For, in order to recover the mine, it is necessary that no more water should flow into the shaft. If the shaft is receiving water at the rate of say ten thousand gallons, it is making a thousand gallons per day more than can be raised to the surface by the aid of ladders and *tanateros*.

Again, in the matter of ventilation, quite as serious difficulty is experienced as in the water problem. Judging from the methods evidently in use in the older and more extended operations, there was but little science employed. Only where the natural conditions were favorable is any attention to ventilating-shafts or tunnels observable. And in these cases no great difference of altitude exists between the points chosen for the entrance and exit of air-currents. Nor is there any evidence of the air having been drawn from, or forced into, the mines by the aid of machinery of any kind. Tunnels of considerable length are found having an access to air sufficient to the carrying on of the work, the ventilation being obtained by means of *lumbreras* (sky-lights), driven from the surface above. The distance between the surface and the tunnel in such cases generally permits the opening of these air-shafts at a moderate expense.

The principle of the Bunsen pump was adopted in Honduras in the first half of the present century, but it does not appear that its use has ever been extended beyond the furnishing of a blast to the primitive native smelting-works.

Furthermore it does not appear that any extensive shafts or tunnels were cut until about the same period. Indeed, so far behind the Mexican was the Honduranian miner, that the device known as the *arraster*, worked by water, was not introduced into the country of the latter until after the discovery, in the year 1747, of the mineral wealth of the district of Yuscaran. Prior to that time, the ore was ground to pulp by hand, or by means of the appliance known by the name of the Chili mill.

One of the most interesting of the petitions presented to the Spanish crown by the miners of Honduras, praying for the improvement of their condition, is dated 1799. Its contents afford a fair illustration of the difficulties under which the mining industry was laboring both before and after that year.

The miners of Yuscaran assembled for the purpose of expressing their dissatisfaction with the existing state of affairs, and of suggesting to the Crown some means for bringing about a revival of the all important industry. Among the hardships complained of, the following are worthy of note :

(*a*) Lack of laborers.

(*b*) Excessive taxation, aggravated by the " red tape " of the local government, which subjected the miners to useless expenditures, involving circumstances entirely without benefit to themselves and very harrassing.

(*c*) They complain, further, of the refusal of the local government to give to the *mineral* of Yuscaran some forty-five workmen, who had been drafted for the service of the mines in that district. Also, of being taxed six and a quarter cents per capita a week for the services of workmen never sent to the mines.

(*d*) That foreigners and merchants bringing produce and other wares to the markets of Yuscaran had been so unjustly taxed that they no longer came thither on business expeditions.

(*e*) That the laborers designated for their mines were either sent away to San Salvador or drafted for military service on the north coast.

(*f*) They submit, that these abuses had greatly interfered with the prosperity of mining in the district; so much so, in fact, that of the thirty-five known veins of gold and silver, only a few were being worked at all; and whereas, formerly seventeen native reduction-works had been constantly in operation, at that time but three were active.

They plead, in conclusion, that their difficulties are not owing to failure of ore in the mines, but to the want of workmen.

While the foregoing petition has reference to but one mining district, it tells the story of the others. How limited were the mining operations of the country, is best seen from the stress laid in that petition upon a request for the early delivery of some six hundred pounds of quicksilver, for amalgamating purposes, which indispensable material is therein stated to be in the city of Comayagua.

The historical importance of the above protest is considered sufficient warrant for introducing into these pages a translation of the terms of the petition.

Since the beginning of the present century, and up to about the year 1823, a more decided development seems to have been effected, particularly in the *mineral* of Santa Lucia. It is said that shafts four hundred feet in extent, and tunnels of even greater length, have been driven in this district, as well as in that of Yuscaran. It does not follow, however, nor is it shown in the description of these mines, that the shaft-developments were accompanied by drifts and tunnels of corresponding magnitude, or *vice-versa*. Indeed, the entire absence of machinery for ventilating and pumping, and the lack of knowledge of the principles involved in these problems, show absolutely that depth was habitually sacrificed to development upon the surface-side of the vein, or, where conducted by means of tunnels, that the lateral developments had no depth to correspond.

The slow growth of mining skill, as evidenced in the history of Honduras, may be illustrated by a reference to the efforts of the Rosas to open the celebrated Mina Grande mine of Santa Lucia, by means of a tunnel. They had successfully worked the vein of this mine from the top of a mountain, also called Mina Grande, to a depth of four hundred feet. There, however, the work terminated; not because of a failure of the ore-body, but simply for want of ventilation. At that depth, they were still some eight hundred feet above the natural water-level, the Rio Chiquito, no water being found upon the Mina Grande mountain.

Unable to cope with the difficulty thus presented, the Rosas imported a number of Mexicans for the purpose of driving a tunnel from a point six hundred feet below the old works. This fact indicates that the Mexican miner was esteemed by the Honduranian as his superior, if not indeed as a master of his trade.

The work of the Mexicans upon the Mina Grande, however, proved a failure. The tunnel, still in existence, was driven in a semi-circle, missing the vein altogether. No explanation can be given for this apparent stupidity, except that the miners had no instrument by which to ascertain or to regulate the direction of their work.

Another tunnel, still in existence, known as the *Gatal*, was driven at the same time. The inaccuracy of the work, the expense, or the political entanglements of the owners, or all three causes combined, led the Rosas to abandon, about the year 1823, this and all other operations in the Santa Lucia district. From that time until the present year, little has been done towards the recovery of these once famous properties.

What has been said of the Santa Lucia applies as well to the other mining districts of the country. It may be added, in general, that the Honduranian does better work as a smelter and gleaner than as a miner.

Every mineral district of Honduras shows the lack of ability to apply correct principles of ventilation, or to handle large quantities of water economically and effectively. In all the mines opened by the Spaniards, the veins were attacked from the highest available point, and worked from that point as long as possible. When a depth was reached too great to permit of bringing to the surface the poorer grades of ore, the work was discontinued, except so far as necessary in order to get at the richer portions of the vein. When ore was mined at a considerable depth, the richer portions were selected and the poorer portions used as material for "filling" or "backing" elsewhere in the subterranean work.

The Spanish miners of Honduras were scarcely unaware of the advantage to be gained by attacking the mine according to well-defined plans, involving the driving of tunnels, shafts, adits, vent-holes, etc. But it was beyond the reach of their resources to provide the requisite amount of expenditure and of labor.

In order to a fuller understanding of the situation at the present time, some description of native methods of mining will not be amiss here.

The Honduranian, though an excellent judge of the quality of ore as revealed by natural indications, still, as a miner, has not improved to any degree upon the skill of his predecessors. In opening a mine, his first efforts are directed to digging a shaft, called in mining vernacular a *poso*. To meet the requirements of the mining code of the country, this *poso* had a depth, formerly of fifteen *varas*, now reduced to five *varas* or about fifteen feet. The next step is an excavation of similar extent upon the vein. This work determines the course and dip of the ore-body, besides other characteristics. Instead of carrying out, from this stage of proceedings, a reasonable plan of development, the *poso* and " gallery" system is persisted in until the interior of the mine assumes the form of a gigantic stairway. This work is extended by "gougings," so that the excavation loses its proper character as a mining shaft.

As to the necessity of supporting this interior work by means of artificial props, no skill has been displayed by the native workman. Such plans as he had were limited simply to one day's requirements. No work appears to have been done with a view to facilitating future operations. The timbers set up within the excavation were intended to provide only for immediate needs. If they outlasted the occasion, it was because the workmen built better than they knew. The fact that native works were frequently lost by cavings shows how poor were the protections provided, and how meagre the knowledge of the principles involved in their construction.

Nearly all the veins of Honduras carry pay-streaks, the extent and richness of which vary as the work progresses. The ore of this pay-streak, called *metal*, is the object sought by the native miner. His methods of mining and smelting demand an ore of greater value than sixty dollars per ton to enable him to make a profit.

The *metal*, so called, usually yields this high grade of ore, while the vein-material, or *brosa*, is carefully selected and added to that from the pay-streak. When a considerable depth has been reached, the selection of these ores is made within the mine, as it would not pay to lift to the surface, in the manner already described, all the vein-material "thrown down" by the workmen. This selected ore is treated in one of three methods, according to its character.

The smelting-works at present in operation among native miners are capable of treating a small amount of ore at a time. This work, although performed by crude machinery, is nevertheless well done. The chief hindrances are the primitive form of the blast itself, and the difficulty of obtaining iron for purposes of separation. In the better-constructed native furnaces, the blast is obtained by driving a column of water down an upright cylinder, according to what is known as the Bunsen principle. In ordinary smelting, it is obtained by means of the common hand-bellows.

Iron of superior quality is found in some parts of Honduras, notably in Algeteca, where the magnetic ore is exceptionally pure. This district of Algeteca is the

NATIVE CONCENTRATING VATS.

property of the Central American Syndicate Company, and thus far no developments have been begun upon it. It is because no one has yet undertaken to open up the iron districts that that great industry remains undeveloped. Hence arises the difficulty experienced by natives in obtaining iron for reduction purposes. The fact is that operations in this department of the work are not carried on with sufficient regularity to create a steady demand for iron. The workmen are content to make use of the oxides found there in abundance, and to reap the benefit of such portion thereof as is converted into metallic iron by the moisture present when the ore is introduced into the smelter.

For refractory ores, the *patio* system is commonly used, while in the very few improved reduction-works, the German barrel-system prevails. By the *patio* or yard-system is meant the mixing with quicksilver of a mass of pulverized ore. This mixing is done in a yard having a hard level floor. Here the ore is gathered into heaps, or *montones*, of about twenty-four hundred pounds each. At stated intervals these heaps are re-mixed, in order that the mercury may come in contact with all the precious metal which the ore contains. A period of six weeks is ordinarily required for the complete amalgamation of the ores of one such heap.

Within the past two or three years a change has come over the aspect of affairs. A road-way, lately built, from the Pacific Coast port of Amapala into the heart of the silver belt, has solved the problem of the introduction of heavy machinery, a thing heretofore impracticable. Several companies are now engaged in developing various properties in accordance with modern principles of mining. From the investigations made by these companies, it appears that the abandoned mines, now the field of their operations, are still very valuable properties. It appears that the amount of ore extracted from them by the Spaniards formed but a small proportion of the original wealth of the mines ; and to-day the only obstacles to rapid development are the indolence of the people and their inability to cope with the difficulties of the situation.

Under the advantageous conditions now prevailing in the mining districts, laborers in plenty can be had, at wages varying from thirty-seven and a half cents to one dollar per day. These workmen, under proper guidance, are as efficient as any, besides being conscientious, tractable and sober. With the new systems of ventilation and pumping lately introduced into that country, mines that were effectually sealed under the old *régime* give promise of large returns to capital invested under intelligent and scientific management. Careful examination of these claims has proved that the mining industry of Honduras, in giving large yields of wealth in the past, has only indicated the present possibilities of its immense resources. The conclusion is believed to be incontrovertibly established, that no country of the world at present contains, in proportion to its area, so great an extent of valuable mining territory, and so rich a deposit of the precious metals.

CHAPTER VIII.

GENERAL INFORMATION.

OF the many enterprises started within the past ten years in Honduras, little is known by the public, owing to the fact that few travelers make a general tour of the country; and information to be obtained from the citizen, is generally confined to works located in the neighborhood in which he lives.

The Opal Mines are near the town of Erandique, in the department of Gracias, and the principal ones are worked by Messrs. Peacock and Burdet; from these mines many of the finest gems find their way into the American and European markets.

Mention may be made of an active business firm at present engaged in a profitable trade in fruits on the north coast near Puerto Cortez, dealing principally in bananas and cocoanuts. From the same point another company exports large quantities of mahogany, cut from the banks of the Chamillicon river. Both these concerns are managed by energetic Americans, hailing from the State of Michigan. Other fruit-growers along the coast ship immense loads of fruit by steamers and by sailing vessels to New Orleans.

The river Aguan is now being opened by a company formed in New York for the purpose of clearing it of obstructions and exporting fruits and woods from its banks. The government has conceded certain adjacent territory to this company in consideration of the improvements thus undertaken.

With regard to the railway now in operation between Puerto Cortez on the Atlantic coast and the town of San Pedro Sula, thirty-eight miles inland, it may be stated that the road-bed is already laid thirty miles further into the interior. The repairs now projected will extend the railway thus far, making Portorillos the inland terminus. This improvement, obviously one of considerable importance to the commercial progress of the country, has been undertaken by General Kraft, Commandante of the department of Santa Barbara, who now holds the road under contract with the government for the next twenty years. The government proposes to augment this improvement by the construction of a wagon-road from Portorillos to Tegucigalpa, the present capital and chief city of the country. The route is considered fully as practicable as that of the road already successfully completed on the west coast from San Lorenzo to Tegucigalpa.†

Other recent projects for the building of rail-roads in Honduras include a plan, already partly carried out, for a rail-road from Truxillo to Puerto Cortez, in the interest of the fruit-raisers along the north coast ; also a road through Olancho from Truxillo to Tegucigalpa, the surveying of which, during the past year, was followed by a favorable report ; and a road to Juticalpa from a point on the Segovia river in the mining district, to be reached by a line of steamers on the river.

There is at the present writing a company of French miners on the north coast, working the old Santa Cruz gold mines in the department of Santa Barbara. Their works are located about thirty miles southwest of San Pedro Sula. This company, which with wise and judicious management has proved most successful, spent a large sum in preparing the mines for working before putting in machinery. This

† See Chapter IV., p. 40, and p. 84.

plan has enabled them to export bullion steadily for about two years, or since their machinery was put in operation.

The only other work of any importance on the north coast, is a hydraulic mine on the boundary line of Guatemala, said to be paying handsomely. In Santa Barbara the mining districts are unoccupied. In Olancho the natives are engaged in mining operations on a small scale.

The main reasons for the difficulty attending *placer* mining in Olancho are the distance from which water must be brought for hydraulic work, and the necessity for its being brought to the mines under pressure in order to displace large quantities of gravel at small expense. There are undoubtedly, in spite of long continued native exertions to extract the precious metals lying nearest to hand, a good many square leagues of rich deposits, chiefly in the hilly country on the Guayape river, awaiting improved methods of hydraulic mining. In addition to the obstacles already mentioned, Olancho is a valley country with a higher temperature than that of the Pacific slope, and not so well adapted to the requirements of health. Transportation there is still very expensive, owing to the fact that no wagon-roads have as yet been built into the interior. The cattle interests in Olancho form its most important and most profitable industry. The ores of Paraiso and Tegucigalpa include free-milling gold ores, refractory gold ores, and silver ores in almost every variety.

A good example of a prosperous mining property is the one known as the Rosario. This was opened upon an abandoned working. An opening was found to have been made by the old Spanish miners to a depth of about two hundred and fifty feet. The old vein had an average width of about three feet. A new deposit of the same vein is now being worked further up the hillside. This latter ore body was discovered in building a road-way up the mountain from the old working. This continuation of the vein is of the kind known as the true fissure and is said to widen in places to twelve or fifteen feet. There is no doubt that the exceptional success of the Rosario mine is largely due to its advantageous situation as regards the question of water. Probably no mine in the whole country is superior to it in this respect.

The Animas mine at the Valle de los Angeles has been profitably worked for a period of eighteen years, the owners being content with the moderate results obtained by means of the crude native smelting processes. About ninety dollars per ton is said to be the average value of silver ore taken from this mine, the vein varying in width from fourteen inches to seven feet, with an average thickness of three feet.

One of the commonest difficulties attending the native reduction process, described in the preceding chapter, is the lack of readily available lead and fuel. Wherever lead and fuel can be easily obtained, smelting may be successfully carried on, with a profit only measurable by the character of the mechanical appliances employed in the process of reduction. Aside from the Yuscaran, Rosario and Santa Lucia Companies, the Pacific slope has been thus far almost untouched by mining operations.

With regard to the attitude of General Bogran, the President of Honduras, towards the mining enterprises recently undertaken there by American companies, that gentleman distinctly states that in his opinion this is the most important work yet begun in that country. He appreciates highly the benefits afforded by the recent practical results obtained in the building of the wagon-road, making a highway from the Pacific Ocean to the capital city and to the principal mining districts. He pledges his government to furnish promptly every needed assistance in these desirable developments. For instance, all citizens being, by the law of the land, soldiers under the requisition of the govern-

GENERAL INFORMATION. 77

ment, and liable for military duty whenever called upon by
the authorities, he himself proposed to excuse from military
service the citizens of the mining district, and henceforth
not to draw from their number any soldiers, except in case
of some political crisis of paramount importance.

There is a government commission now engaged in ex-
amining the mineral and agricultural interests of the whole
country, and in preparing accurate maps, so as to enable
the government to publish, both in English and in Spanish,
a clear report of the resources of Honduras, as a standard
of accurate and authoritative information upon the subject.

As a matter of historical fact, very important in its
bearings upon the present situation, it must be remembered
that when Honduras obtained its political independence,
the Spaniards who had owned and worked the mines
returned, with but few exceptions, to their native country.
Anarchy reigned thereafter for a long period. Governments lasted but a few months at a time. The people had
been slaves, and lacked both the means and the intelligence
necessary to successful and continued industry. Gradually,
however, as affairs settled themselves, and certain of the
more energetic people acquired property, the business of
mining was revived under a crude system of operations.

For a long time the laws prohibited foreigners from
coming in and acquiring property, and all outsiders were
prevented from working at the mines. Not indeed until
the presidency of Mr. Soto, the predecessor of President
Bogran, was this obnoxious law abrogated, and the same
privileges of citizenship freely accorded to resident foreigners as were enjoyed by the natives. The former prejudice
against foreigners has since happily disappeared altogether.
Americans coming to Honduras on business are especially
looked upon with favor and received with welcome.

The fact that the Indians in Honduras were actually
slaves prior to the separation from Spain, was one reason

why they did not look favorably upon the coming of Americans to their country. They feared that an influx of Americans would issue in a return of the old hated conditions of servitude from which they were now freed. But as soon as they understood that slavery had been abolished in the United States, their sentiments towards Americans were radically changed. Then it was that the new laws were enacted.

For a number of years past, the so-called "liberal party" has been in power in Honduras. This party favors the continuance of the existing separation of Church and State. The representatives of the opposite idea form the conservative party, which is numerically very small, and yearly growing smaller.

In Spanish American countries generally, the president is a dictator. It is commonly conceded by the people that the country and its resources are at his disposal. While, therefore, there is a Congress, composed of representatives from the different parts of the country, having nominally the controlling voice in state matters, and whose duty it is to confirm, or withhold confirmation from, the acts of the president and his ministers, yet when this congress is not in session, the president has full power delegated to him to act upon his own judgment in the affairs of the government. The Congress itself, for that matter, is made up of men that are, almost without exception, in perfect accord with him on all questions. Care is observed to fulfill all legal requirements in any state action, to the extent of considerable red-tape, even in unimportant matters.

At the times of presidential election, every resident, whether foreign or native, is entitled to cast his vote at the polls. Votes are cast directly for the president, instead of for electors. The number of candidates is limited, however, and there is usually no special hindrance to the re-election of the president, if he desires to continue in office.

GENERAL INFORMATION.   79

Honduras is emphatically a mining country. Its people have been devoted chiefly to this industry, and it has produced, among the states of Central America, by far the greatest amount of gold and silver exported in time past. There has never been as much attention paid to agricultural pursuits in Honduras as in the neighboring states. In Guatemala, for instance, agriculture is the principal industry. Only within the past fifteen years have any large plantations been cultivated in Honduras, in spite of its vast resources. We must except, however, the large plantations of cochineal cactus that once existed in this country, when the cochineal trade was one of the principal industries, but this is now a thing of the past, owing to the introduction of aniline dyes.

Health may be preserved by observing hygienic rules, and little fear of malaria need be experienced by the traveler, excepting when passing through the low lands extending thirty miles back-from the coast, where the temperature ranges from 90° to 100° Fahrenheit. At midday there the heat is as oppressive as in our midsummer. In the dry season the liability to malaria may be evaded by keeping within doors at night ; and at other seasons there is little risk of this kind, since this low district may be crossed and the high ground reached in a single day's riding.

The principal mining districts are all situated in the pine region among the hills, where malaria is unknown. Diseases of a typhoid character, common in the United States, do not exist in Honduras. Pulmonary trouble is rarely heard of. Cases of small-pox occur at times, making it prudent for the foreign visitor to guard himself by vaccination, previously to taking up any prolonged residence there.

In the pine region the warmest days of summer do not go beyond a temperature of 75° to 90°. The nights are always cool, and comfort requires at time the addition of a second blanket for covering. In short the climate of the mining country is as equable and healthful as that of Pennsylvania or New York.

The native population of Eastern Honduras are all *Ladrinos*, descendants of the Spaniards and of the aboriginal Indian stock. There are but few whites in this section. The other half of the country, namely, the southern and western slope to the Pacific, is the more thickly populated portion, yet even here the number does not exceed two hundred thousand souls. Prior to the decadence of the mining industry, the population was much greater; but as business activity diminished with the departure of the Spanish employers of labor, the inhabitants of Honduras drifted into the neighboring states.

There is, in fact, but little difference between the natives of Honduras and those of Guatemala. In their race-characteristics, political and religious institutions, and habits of life, they are much alike. The Central American people are to a certain extent homogeneous, distinct in many important features from the people of Mexico on the one hand, or of South America on the other. This is largely due to the character of the aboriginal Indians of Central America, who were notably superior to the savage tribes from which, for example, the modern desperadoes of Mexico are descended. This may help to explain why in many parts of Mexico military escorts are needed to-day for travelers or for the transportation of valuable goods, whereas in Central America brigandage is unknown, travelers are never molested, and merchandise of every sort may be carried over the least-frequented roads without fear.

Again, in the case of the Central American states, there is no ground for political animosity toward our own country. On the contrary, the inhabitants evince a strong desire to imitate our institutions, the surest token of admiration and general good feeling.

There is much room for improvement as regards industrial development and business enterprise throughout these states. Considerable advancement has been made within the past ten years, but chiefly in Guatemala, and least of all in Honduras. This is true in respect not only to roads and railways, but also to the introduction of improved machinery for mining and manufacturing purposes and for the preparation of agricultural products for foreign markets.

For example, at Yuscaran, a saw-mill lately put in operation is the second of its kind ever taken into the interior of Honduras. What little machinery for mining purposes is now there was brought thither within the past three years. The crudely-fashioned box-bellows used by the native blacksmith has only just been replaced in some instances by the patent portable forge. The very primitive contrivance for smelting the ores has been elsewhere described, as well as the tedious method of lifting the *metal* from the shafts.† The processes still in vogue, even at the richest veins, scarcely permit of treating more than a single ton of selected material per day. Yet even thus good results are constantly attained. So much so, that there can be no doubt that with proper mechanical equipment for hoisting, pumping and ventilating, under intelligent management, supplemented by modern appliances for smelting, these mines would be immensely productive. All the natural requisites for successful working are available on the ground. In many localities there is ample water-power even in the dry season. Abundance of pine and oak provides material for building, shaft-work, fuel and charcoal. The necessary fluxes for smelting, to wit, silica, limestone, and iron, are either present in the ores or obtainable from deposits near the mines.

To judge from the official records on file among the state and municipal archives of Honduras, the amount of

† See Chapter VII., pp. 63, 70.

wealth produced by the mines in years gone by was simply enormous. For example, the Guayabillas mine alone, on the basis of "the king's fifth," a tribute to the crown always scrupulously exacted, paid taxes on its yield, from 1813 to 1817 inclusive, a period of five years, of no less than $400,000. This shows a production for that brief period of not less than two million dollars. Again, in 1845, before the last abandonment of this mine, when the property was taken into the courts on a suit for damages, the amount of damages claimed, as the records show, was one million dollars. Similar fragments of history might be cited with reference to other mines of this and the neighboring districts.

That the precious metals could not have been exhausted by former workers in the mines is proven beyond the shadow of a doubt. First, because mining was being carried on in 1820, until the separation from Spain took place ; at which time, as elsewhere explained,† the laborers of the country were scattered by the disorganization of society and of industry resulting from this radical political change. Secondly, because the shafts and tunnels could not be continued very far below water-level, owing to the lack of adequate means for pumping out water. Thirdly, because those who were engaged in mining operations were utterly ignorant of the principles involved in the matter of artificial ventilation. Fourthly, because none of the old mines thus far examined show evidence of having been worked beyond a depth of four hundred feet ; nor has any miner been found in the country who knows of any mines showing greater extent of operation than this, either by shaft or by tunnel.

There are certain mistaken popular impressions in our country with regard to Central America, which ought to be corrected, if only out of respect to the general intelligence

† See Chapters II. and IV., pp. 20, 37.

GENERAL INFORMATION.   83

of well-informed people. The common belief even of educated persons seems to be that the country in question, although not further removed from the city of Washington than is the state of Colorado, is a region of inaccessible mountain plateaus, unfit for human habitation by reason of frequent volcanic eruptions and shocks of earthquake, or a region of endless swamp and jungle infested by venomous reptiles and insects, and poisoned by a malarial atmosphere of the most dense and deadly kind. So vivid is the uninformed imagination!

It is true that there are volcanoes in Central America, but they are more interesting than dangerous. As to earthquakes, rarely has a loss of life been occasioned by them. In Honduras, there are no active volcanoes, and the latest earthquake which disturbed even large buildings occurred in 1811, when the roof of the cathedral at Tegucigalpa was slightly damaged.

In the pine region, as we have said, there is no malaria. This timber-land embraces hundreds of square miles of territory, and includes within its boundaries most of the richest mineral districts of the country. As to reptiles, they are no more prevalent than in the mountains of New York state; and insects receive as little attention as in our own Southern states.

The attention of energetic men of capital has of late been especially directed toward the great undeveloped resources of this accessible field for investment. Such an impetus has been given to the mining interests alone that a considerable emigration of skilled American mechanics, miners, and metallurgists only awaits the determined energy of enterprising capitalists to lead the way. Foreign residents there acknowledge that American interests must soon dominate all others in the country. Most of the railways now in operation in Central America were built by American capitalists, and are owned in part and operated entirely by Americans. The time-honored methods of the English and German traders will not be likely to hold out long against Yankee push and ingenuity. Many articles of American manufacture already find entrance into Honduras from the port of New Orleans, and this branch of trade is on the increase.

Meanwhile, the governments of those states court a continuance of the protection afforded them by the United States against encroachments, territorial or other. A concise statement of the general political situation, as affecting the interests of commerce and as bearing upon international relations, will be found in another chapter.† The question of popular education in Central America and its beneficent results and hopeful promise for the future is likewise elsewhere discussed.†

Great injury was doubtless done to Honduras in particular, by the unfortunate failure of the Inter-Oceanic Railway project some years ago. This was a serious setback to the development of the country, besides imposing a heavy burden of debt upon the government. Under the wise administrations of Presidents Soto and Bogran, important progress has been made, however, in the direction of education and general enlightenment. Under President Soto, for example, a college was established at Tegucigalpa, and common schools in all the towns of the state. All the principal points in Honduras have telegraphic communication. There is a good mail service, and a well-built and admirably conducted state hospital. The opening of the new road-way from San Lorenzo‡ was begun by President Soto and assisted to its completion by President Bogran. This is undoubtedly the beginning of what will prove a development of the most important interests of the country, commercial, social, and political. Not only is the government itself committed to the persistent encouragement of

† See Chapter IV. ‡ See page 40.

all honest efforts for such improvement, but every influential citizen, without a single notable exception, is in active personal sympathy with the movement now in progress.

It is the lack of facilities for the exchange of the products of different districts, and for the transfer of goods between the coast and the interior, that has been up to the present time the chief hindrance to systematic and permanent growth. There was a time when transportation to and from the capital of Honduras was better provided for, as is evidenced by the fine stone bridge, several hundred feet in length, across the Rio Grande. This structure was undoubtedly erected when the affairs of the adjacent mines were in a flourishing condition. That period of her history from a century to a century and a half ago, must have seen Tegucigalpa the home of a numerous and thriving population, four or five fold greater than at present.

The new road-way follows the line of the old original thoroughfare. It makes the same fords, has much lighter grades, and is of greater width. As to the means of carriage at present employed, the rude native two-wheeled carts and cargo-mules are made to answer all immediate needs. The carts are drawn by cattle trained to drag them over rough rocky paths and up steep ascents. Formerly about a month's time was occupied in transporting a few hundred pounds of freight from the coast to Yuscaran. With the newly finished road-way, American four-wheeled wagons drawn by mules will be able to take in loads six times as heavy at no greater expense. Cargo-mules carry about two hundred and fifty pounds, if the load be well balanced, at a cost of three cents per pound, reaching Yuscaran in ten days' time. Travelers to the interior ride on mule-back, as described in our introductory chapter.†

Primitive conditions such as these will account for the fact that the mineral wealth of the country still awaits

† See pp. 15, 16.

development. If the official records of production in times past are to be taken as they stand, it is remarkable that methods so rude and facilities so meagre should have resulted so handsomely. This points unerringly to the rich quality of the ores that have already been turned into bullion. While, at the same time, the comparative inaccessibility of the mining districts, before the building of the new road, explains the fact that so little is commonly known of the country, both at home and abroad.

It must be granted that in Honduras, as in other countries in the same latitude, business is generally transacted by slower methods and on a smaller scale than in communities like our own, where the rule is "quick sales and small profits," and where time is naturally held to be of greater value. The business man of Central America, with few exceptions, is never in a hurry. As matters stand, there is little or no competition, hence his business will wait his convenience. Time, however, will bring about a change in this respect. At Guatemala, there are two banks issuing notes which circulate at par. The government of Guatemala issues legal tender notes and gold coin, and business in that state is generally conducted on a broader basis than in Honduras, where loans command two per cent. a month, exchanges are made exclusively in silver, and United States money and New York drafts have a premium of twenty-five per cent.

As to the cost of labor in Honduras, the working people are rather more intelligent and active than those of the neighboring states, and demand better wages in consequence. In the coffee and sugar plantations of Guatemala, for example, a laborer receives twenty-five cents per day, while in Honduras day labor of all kinds is paid thirty-seven and a half cents and upwards. The laborer is by nature more clever and apt than he is industrious. Less exertion is required than among ourselves in order to provide for his

GENERAL INFORMATION. 87

wants and the wants of his family. These wants are extremely simple. The coming in of foreigners will improve the conditions of living, and increase the incentives to industry and thrift on the part of the working people. When engaged in tasks to which they are accustomed, they show great powers of endurance. For instance, in all the mining districts, workmen are to be found who are familiar with mining operations as they have been carried on, and these men, under firm and considerate direction, make good laborers.

In order to encourage the re-opening of mines under foreign management, the mining laws have been amended, so that miners may be regularly articled to such company for six months or more. When articled, they are exempt from military service, and are only required to attend militia drills and musters once a month. As a matter of fact, there are now more applicants for articulation and for employment than there are places to be filled.

For example, in building the new road-way from the coast, laborers were furnished to the contractors on requisition from the government, and paid fifty to seventy-five cents per day, foremen one dollar and a quarter. At the mines, the laborer receives fifty cents per day, or more, according to his work, mechanics one dollar and a quarter. Levels and tunnels are driven by contract at a cost of from ten to fifteen dollars per yard, the companies furnishing timber and supplies. These are the rates prevailing at the present time.

It may be seen from what has been stated, that there is an abundance of workmen already in the country, and that the opportunities there afforded for the profitable employment of capital in mining, fruit-raising, mahogany-cutting, and other industries, are perhaps unequalled in the world. There is every reason to believe that the Spaniards of former days and the native miners of the present have gotten practically most of the precious metal that could be obtained with the help of their primitive appliances ; and that it is only necessary to go there with proper hoisting, pumping and ventilating machinery, supplemented by modern methods for rapidly reducing the ores, in order to reap the rich harvests of those mountain-sides, now no longer beyond the reach of cultivation by a judicious combination of science, energy and capital.

ERECTING MACHINERY AT GUAYABILLAS MINE, YUSCARAN.

CHAPTER IX.

YUSCARAN.

DURING the earlier years of the administration of President Soto, while he was endeavoring to turn to account the great natural resources of the country by bringing them to the favorable notice of foreign capitalists, many persons of scant influence at home made application for valuable concessions in connection with mines, railroads, and other industries, representing themselves as men of means and claiming to have a wealthy constituency behind them. As a result, the government was considerably hampered in its efforts to open up its resources to the commerce of the nations. It was discovered that these individuals were unworthy of the privileges that were accorded to them, since they did not hesitate to try and discount their valuable concessions in foreign markets at a paltry figure, caring but little whether or not any actual use was made of them.

The president realized that the most effective means to the desired end was the reorganization of the dormant mining interests. He was convinced that this would bring about a revival of trade in the rich products of the republic, by reason of the consequent general infusion of new life and modern methods, together with the requisite energy for carrying them into practical execution. For the sake of this result he was willing to enter into business negotiations on a liberal scale, and to make arrangements with reliable parties upon such a broad and generous plan, that Americans, especially, perceiving the immense natural advantages ready to hand, would appreciate the wise and intelligent co-operation of the chief executive, and be prompt to supply the necessary capital and energy for the proper development and success of these projects for national and industrial advancement.

It seemed probable that, unless such systematic organization could be effected, the country might continue at the mercy of mere adventurers, without the permanent accomplishment of anything in the way of actual progress, commensurate with the well-known value of the interests involved.

It is now but a few years since this idea took practical shape in the formation of a syndicate of American gentlemen, with the primary object of the opening up of the mining districts, the promotion of railroad projects and of such other commercial operations, as should present themselves in the gradual development of the abundant resources of the country. This organization comprises a number of energetic and responsible citizens of the United States, whose names are a guaranty of good faith. The intention was that this corporation of individuals should not only actively undertake certain business operations, but also be related in an advisory capacity to the local government. By this means the latter, being able to obtain trustworthy information as to the standing and integrity of parties proposing to engage in business in that country, could be guided intelligently in its negotiations with them. Hence, in order to constitute the aforesaid organization a reliable agent for all such as might in future turn to Honduras as a field for investment, extraordinary powers and especial privileges were granted, with permission to extend the benefits and advantages thereof to individuals or corporations undertaking work in that quarter under these auspices.

The concessions conferred under date of April 30, 1882, embodied a practical control of the most important mining departments of the country. A large number of mines were conceded in fee-simple, together with the exclusive right to erect custom reduction-works, the right of patent on all machinery introduced into these districts under control of the organization, exemption from taxation of every kind, whether external or internal, exemption from duties upon all articles or materials imported for use in the mines or for the needs of the operatives. It will be seen from the liberal terms of this concession that all persons or companies doing business under these conditions, are ensured entire immunity from the vexations that commonly beset strangers in Spanish-American countries, when engaged in commercial transactions under the general laws.

Immediately after the formation of the syndicate as stated above, a search was made in the archives of the country for definite proofs as to the value and prospects of the mining industry. The preceding chapters† of this book afford abundant evidence of the satisfactory results of that investigation. Since that time, several American companies have begun operations in Honduras in the different departments of the State ; principally at Yuscaran in the department of Paraiso. Here are located the works of both American and native mining companies, now engaged in developing their mines by means of shafts and tunnels. On some of the shafts there are "plants" of hoisting, pumping, and ventilating machinery equal to any of a like character in the mining districts of the United States.

The little town of Yuscaran, which five years ago was at a very low ebb, only sparsely populated and enjoying but a limited trade, is already become a thriving business center, showing even greater activity in this respect than the capital city of the republic. Its merchants are prosperous, the people busily employed at the various mills, the steam-whistles are daily heard, machine shops

† See especially Chapters VI., VII. and VIII.

and saw-mills are at work, and the general aspect is one of enterprise, progress, and thrift. In the mines themselves, adjacent to the town, the development thus far attained (March, 1887) gives every encouragement to the owners, and the time is believed to be near at hand when the output of bullion from this quarter of the mining industry will be an important factor in the general output of the world.

Some more extended description of this particular center of growing commercial activity cannot fail to be of interest to the reader whose attention has been drawn to a region heretofore comparatively unknown and yet replete with picturesque and attractive features.

Nestling among the fertile slopes of the Plata and Santa Elena mountains, Yuscaran affords one of the most charming pictures to be found in all Honduras. The traveler, winding his way through the neighboring country, receives no intimation of his proximity to the town, guarded as it is by huge mountains rising on every side, until, ascending a prominent spire of one of them, he beholds the village but a few hundred yards distant,—its low white houses and sentinel-like church standing out with startling distinctness against the green background of the wooded hills beyond.

Looking away eastward, across broad rolling cattle-plains, plentifully watered by rivers and streams, one sees, only a few miles off, the boundaries of Nicaragua; while to the north, south, and west, the stately mountains rise to a height of three thousand feet above the little town. Yuscaran itself, having an altitude of 3,250 feet above the level of the sea, enjoys a magnificently clear and bracing atmosphere. Indeed, the climatic advantages possessed by the residents of the town are without parallel in any part of the republic.

The average temperature the year round is 72°, and yet, owing to the valleys and lowlands on one side and the mountains on the other, a difference of 20° either way may be experienced within an hour's ride from the *plaza* of the town. Under these conditions, together with the advantages of a soil remarkably rich and fertile, it will readily be understood that fruits, vegetables, and cereals of almost every kind, can be successfully raised.

In fact, the market place of Yuscaran affords a produce exchange for the entire department of Paraiso ; all the towns from the great Indian settlement of Texiquot to Danli, the center of the coffee district, sending every week their several products thither. On the broad plains round about this important town, not only the finest coffee in all Central America is cultivated, but also a superior quality of sugar-cane, in such quantities that the *aguadiente*, or native rum, distilled therefrom, is sufficient to supply the demand of the entire department of Paraiso, and that of the department of Tegucigalpa as well.

The natural agricultural advantages of Yuscaran, although contributing to its advancement, do not furnish the principal reasons for the importance it has attained. It is true that these exceptional facilities for making the life of the resident agreeable, have raised Yuscaran to its present position of popularity. Still, even these attractions would scarcely have won appreciation, had it not been for the heavy veins of the precious metals known to exist in the adjacent mountains, and only awaiting development in order to make this district one of the foremost among the mining countries of our western continent.

During the early part of the eighteenth century Yuscaran consisted merely of a group of Indian huts and ranch-buildings, serving as shelter to the workmen employed in the mines of Portorillas, during their journeys by this route to Tegucigalpa and Nicaragua. The inhabitants of the place had long envied the people of Portorillas, where the rich mines gave employment to many.

The latter in turn looked down upon the humble village of Yuscaran, thinking it afforded a good *posada*, or halting-place, and nothing more.

The vast mineral wealth of Yuscaran was discovered in a somewhat remarkable manner, which it will be interesting to recount here. To Juan Calvo belongs the honor of having brought this store of long-hidden treasure into the light of day. His name has long been held in notable esteem by the good people of the town ; among the merchants, on account of his fortunate discovery, and among the younger men, on account of the brilliant reputation established and sustained by him, as a spendthrift and a gambler.

Calvo once, while riding over a rocky pass in the Plata mountains, in the year 1747, was endeavoring to make up time lost on his trip to Portorillas, and with this intent took an unused path down the steep mountain-side. This path, being somewhat rough and difficult, afforded but an insecure foothold, in many places, for the mule on which he rode. Midway down the steep incline the beast stumbled, threw his rider, and rolled over and over down the rocky slope. Calvo, in clambering down to recover the animal, noticed that a small spur of rock on the hill-side glistened brightly in the sun. Picking up a piece of rock that seemed to be newly broken, he discovered that it had been dislodged from the spur by the mule's striking against it in his fall. To Calvo's great amazement he saw that what he had supposed to be only common rock was in reality silver ore of remarkable purity and richness.

Alarmed by this unexpected discovery, he left the spot, after carefully covering up all traces of his adventure. Returning thither after the lapse of some days, with the help of a few very primitive tools, he began work, secretly and at night, upon the vein he had discovered, grinding and smelting the results of his labors in the most remote

and retired spots he could find. In a few weeks Calvo was
observed to be in possession of large amounts of money,
which he scattered right royally among his friends. He
gave a ball at which he was known to have lost in gaming
a quantity of silver which to his companions represented a
small fortune. Upon being bantered with the remark that
his purse would not long stand such drains upon it, his
reply was that he could always command all the money he
might need. This unguarded statement excited suspicion.
He was followed, and his secret became known.

At once miners began to gather from all sides to work
the vein he had discovered. Inasmuch as Calvo had taken
no open steps to secure the property, it was rapidly taken up
by the new comers, who were soon busily at work in large
numbers developing the mine, which now acquired the
name of the Quemasones. Excitement ran high. Discovery followed discovery. New veins were opened, and
the work was pushed forward with much energy. The rich
mine-owners of Portorillas, convinced of the superior excellence of the deposits of silver ore at Yuscaran, moved
thither almost in a body, bringing with them their slaves
and all their tools and appliances. From this influx of
population, and on account of the constantly increasing
importance of the results of their work in the mines,
Yuscaran experienced great changes in a short time.
Houses were built, streets were laid out and paved, and a
large cathedral was erected.

It seemed indeed as if nature herself had provided
Yuscaran with every requisite for the growth of a prosperous mining-camp. The three rivers furnished abundant water-power. The neighboring hills were covered
with thick forests of pine, while in the valleys flourished
many varieties of the harder and more valuable woods.
Under the impulse derived in part from these natural advantages, the work of extracting the ores progressed rapidly.

A list of the chief veins discovered and worked during this period will evidence the activity displayed. Among the most important are the Quemasones or Communidad, Guayabillas, Sacramento, San Juan Monserrat, Santa Elena, Tomagas, Suyate, Jesus, California, San Miguel, Iguanas, Capiro, Platero, and the Veta Grande, or "Great Vein."

Along the river-banks the mine-owners now began to build reduction works. These consisted of *arrastras*, tanks, barrels, *patois*, and smelters. As the mining industry grew in importance, other mills were erected, until at length along the Rio Grande, the Rio de la Auroria, and the Rio de los Ingenios, could be seen many substantial buildings, in which large quantities of ore were treated by machinery.

These metallurgists of by-gone days were not skilled in the art of saving the precious metals, in the processes to which they were accustomed. They had little idea of so systematizing their work that the greatest possible results might be obtained at the smallest expense of time and labor. By their crude methods, the ores were handled over and over again. Quicksilver, employed in amalgamation, was lost in large quantities, while but little more than half the actual value of the ores was extracted. Under these circumstances, it does not seem strange that the miners of those days preferred to sort their ores at the mine, having it broken up by hand and carefully picked over. In fact, all ores that would not yield, to these methods of treatment, at least sixty dollars per ton, were rejected as too poor to be profitably handled.

In this way the years rolled by. The mine owners grew richer, the town more prosperous and important. Meantime the peons became restive under the bondage to which they had long been subjected. At last, seeing no other hope of deliverance, putting forth an energy born of desperation, the slaves struck a decisive blow for freedom. The bitter conflict which ensued put a stop, for the time being, to the mining industry. As a result, the impoverished people were obliged to turn to agricultural pursuits in order to obtain a livelihood. With the slow return of comparative prosperity, attention was again directed to the hidden treasures of the mountains. Old mines were reopened, and new ones were discovered, when a recurrence of civil dissensions again interrupted the work. Political questions agitated the minds of the different parties and stirred them to action. Revolution swept over the territory, and once more the industries of the country were temporarily paralyzed.

From 1823 to 1876, a period of intermittent disturbances of this character, the mines of the country were worked only after a spasmodic and desultory fashion. What little was done in this direction was mainly carried on at Yuscaran. For example, the great Guayabillas mine was owned and worked by the Englishman Bennett, who brought over from England a large force of Cornish miners, employing them for a number of years and clearing by their assistance hundreds of thousands of dollars.

A single instance may be cited to show this man's estimate of the value of the property. It being impossible at that time, on account of the condition of the roads, to bring in proper pumps to draw the water out of the mines, Bennett resolved to drive a tunnel through the solid rock, at an estimated cost of a million dollars. This work once completed, he would successfully have drained the mine. The sudden death of this energetic individual prevented the accomplishment of his admirable plan, and from that day to this no one has been found with determination sufficient to carry this project into execution. General Xatruch, the last man who worked the Guayabillas mine, obtained possession of the property by inciting a revolution. He removed the pillars of rock from the interior, realizing from the ore of these pillars alone, it is said, more than five hundred thousand dollars.

During the same period, the Quemasones mine was worked by different persons, and with remarkable success. So rich was this particular ore, that five years ago not a pound of it could be found in Yuscaran, every particle having been treated by the natives in their primitive reduction-works. The property was worked to the water-line, when operations ceased for lack of draining facilities. One of the owners of the Quemasones, some twenty-five years ago, took out of the vein a nugget of solid silver weighing no less than two hundred pounds. From this mass of metal an elegant candelabrum was made and hung up in the old church, where it was seen and admired by all comers. At the burning of the cathedral in 1872, this valuable work of art was unfortunately destroyed.

It was now fully realized that further work upon the mines under the old Spanish system was impracticable. Not more than fifty miners were employed in Yuscaran, where in former times work had been found for several hundred men. Hence it is not a matter for surprise that the town itself should have passed through a period of deterioration. The native miners, always considered to be the best in Central America, departed to find work in Nicaragua and Salvador, and Yuscaran, like Rip Van Winkle, sank into a long sleep. *Ingenios*, once the scene of bustling activity, slowly crumbled away. Shafts fell in and were filled up. The town for a time was only one of the many "deserted villages" in Central America. The few inhabitants for the most part knew nothing of the outer world and cared but little. It is true, as aforesaid, that a mere handful of miners were employed, and two or three new veins opened, but this was done in such an unsystematic and primitive way, that the results were not sufficient to cause any material change in the general aspect of things.

When one looks at the Yuscaran of to-day, and recalls what its condition was but six short years ago, the transfor-

mation seems incredible. One who has not been on the
ground cannot realize the improvement that has been
wrought in so brief a period. This change for the better
in the condition of affairs, these giant strides forward in
the direction of a higher civilization, the readiness shown
in appreciating and adopting American ideas, form only the
natural sequence to the work undertaken there by American
business men during the few years just past.

The people, recognizing their need of new blood and
modern methods, have learned to extend a cordial welcome
to those who come among them with the energy necessary
to put these methods into active operation. The companies
represented by the new-comers found a town well-nigh deserted by its able-bodied men, in the midst of a region rich
not only in traditions but in the possibilities of actual available property. None knew better than the inhabitants themselves that the mines could not be successfully worked
without able and intelligent outside aid. But although, dis-
couraged by long familiarity with this state of things, they
had permitted almost everything appertaining to that industry to fall into disuse and decay, they offered to their new-
found friends a hearty and complete co-operation, from the
chief executive down to the humblest citizen. If every tool
needed, and every pound of steel or iron for use in and
about the works, had to be brought from the United States,
willing hands were found in Honduras to put them to use.
It being generally recognized on all hands that the
Americanos had come to stay and to work with them, the
people, one and all, rendered every assistance in their power.
And to this spirit of prompt appreciation is due no small
part of the advancement to be seen to-day at Yuscaran.

Mines have been opened, furnaces, barrels, pans,
smelters, stamp-mills, and other reduction-machinery for
the treatment of both gold and silver ores, have been transported thither and put in place. Saw-mills are in operation,

wagon-roads have been cut, houses and stores built, and work found for a thousand men. There is already a town full of busy, cheerful people ; each one as full of hope and confidence as to the present value and future promise of the mining properties as is the largest stockholder interested therein.

In closing this chapter, some mention should be made of the excellent society to be found at the present time in the town of Yuscaran. The women are gentle, refined, and hospitable ; the men courteous, intelligent, and prompt to assist the stranger by every means at their command. One is made to feel at home at once among them, so cordial and genuine is the welcome uniformly extended. True Spanish hospitality prevails on every hand ; visitors and traders from foreign lands especially receiving freely, from one and all, the greatest consideration.

SAN ANTONIO MINES.

CONCLUDING STATEMENT.

THE Central American Syndicate Company was organized in the city of New York, in 1882, having acquired, as the basis of its operations, special concessions from the government of Honduras, assuring to it the fullest advantages in the wide field for commercial development which that country affords. Experience has already conclusively proved that these particular grants were drafted with exceptional discretion and judgment, their main object being the promotion of business plans involving the development of certain distinct properties, which upon impartial investigation give reasonable promise of abundant success.

The general management of the affairs of the above named company is in the hands of gentlemen who have the advantage of personal experience in Honduras, and who can hence speak from their own knowledge of the favorable opportunities for investment offered at the present time. They enjoy the co-operation of a corps of experts now at work preparing a digest of the resources of that country; men who may be depended upon as sources of direct and reliable information, and whose services the organization unhesitatingly offers to any persons wishing to inquire into the value and prospects of any particular section.

From this statement it is by no means to be inferred that independent projects on the part of American business men will fail to receive a hearty welcome in that country or lack the energetic aid of the resident citizens and tradesmen. No one organization can lay claim to a monopoly of good-will. The government and the people are too thoroughly alive to the possibilities of the situation to omit any effort in encouraging the investment of foreign capital in the various departments of commercial activity.

Emphasis is here given to the combination of advantages offered by this organization; to wit, the experience of its managers, the assistance of its corps of experts, and the promptness of the government and the people to co-operate in furthering those business undertakings which the company endorse. Facilities are now perfected for the furnishing of trustworthy information to all who may desire it, as well as for the securing of such special privileges from the local government as circumstances may require.

www.ingramcontent.com/pod-product-compliance
Lightning Source LLC
Chambersburg PA
CBHW020122170426
43199CB00009B/605